Book

4 of 7

K'Narf Etaguf
II

By:

Steven A. Vaughn

ISBN 978-0-9653330-3-0(E-Book)
ISBN 978-0-9914575-9-5(Soft Cover)
ISBN 979-8-9857125-5-1(Hard Cover)
TXu 1-754-463(L. of C.)

Cover Design by M. Joyce

Dedicated

To my best friend and to my only son who both now sleep waiting for the final resurrection, and to my mother, who has now joined them for gifting me with a love for words. Finally, and with grateful heart, to my wife who patiently listened to each chapter being described and then diligently reviewed them after they had been written.

PROLOGUE

First and foremost, this is a work of fiction. The characters, even though some of them were drawn from the Bible or current events, are still fictitious and none of them reflect negatively on anyone currently living in this century.

Second, the Bible, by its very nature becomes a primary source and, to a limited degree, a central theme when writing about prophetic events. However, I am not a preacher, nor a Sabbath or Sunday school teacher, and have therefore deliberately avoided using this book as a pulpit.

Third, all biblical quotes were drawn from the King James Version.

Enjoy!

Table of Contents

Chapter One

"K'Narf, where are you?" Toby flashed, looking around at the mass confusion caused by the hundreds of people all trying to share the same workspace that was the assembly floor.

"In the gee dunk sneaking a cup of coffee, why?"

"Just got a weird flash from Brent, he wants us to meet him at the Whale. Sounded strange, kinda like he was scared. Didn't even give me a chance to argue about being rather busy at the moment." Toby said, sending the strange flash he'd gotten from Brent to K'Narf so he'd know what he was talking about.

"You're right, that was weird. We don't hear from him for three months and then out of the blue he demands that we drop everything and meet him in the Whale. Have you tried flashing him back?" K'Narf asked, frustration showing in the tone of his voice.

"Yes, he's not answering."

"Mind if we take the Lady?" K'Narf asked, knowing they had no choice but to drop everything and head for the Whale. "T.J.'s already up there." He added, thinking he'd have to flash her and tell her to stay put, that he and Toby were on their way up.

"What's T.J. up to?"

"Flew a Limo up. Leaving a newbie she talked into staying up there and helping Wanda and Rudy. She's setting her up in Wanda's old bedroom. It's been empty since she and Rudy bonded and T.J. thought they could use some help what with the constant flow of newbies going through final orientation these days."

"You ready?" K'Narf asked, as he swallowed the last drop of his coffee.

"No, but this had better be good!" Toby said, the meaning of his words needing no explanation. "Meet you out front." He added, as he stepped over and told Mike he'd be gone for a few days.

"Then give me back my wrench." Mike said, holding out his hand, obviously not happy that his helper was leaving him.

"Yes, of course, here." Toby said, taking the half-inch box-end out of his hip pocket and handing it over. Sorry, don't have a choice, got'ta go." Moments later he was sitting in the passengers seat of K'Narf's Lady as they sped out of the drive and headed for what they knew was a radar safe corridor where it would be safe to lift off.

"You're not going to try and lift off in broad daylight are you?" Toby asked, surprised that K'Narf was obviously preparing the Lady for an immediate lift off.

"Nobody'll see us." K'Narf said, as he finished his computer settings and leaned back. The Lady had changed dramatically since he had first built her. She had onboard computers controlling all her trajectory computations, an atomic battery taken from the downed Snark mother ship and now a new paint job.

"How, I mean..." Toby stammered as they began to lift straight up on a heading for outer space.

"New paint job." K'Narf said, grinning over at Toby. "Glenn and his crew did it over the weekend."

"Okay, give! What's with the new paint? Is this another one of Brent's things?" Toby asked, realizing something new had slipped by his attention.

"Yes and no. Glenn and his team based it on Brent's computer changeable paints he uses on the team vans to disguise their identity, but they speeded up the changes and did some fancy new programming."

"Like what?" Toby asked as he looked out the window at the earth dropping away below them.

"Have you ever seen pictures of a male squid courting a female?" K'Narf asked, grinning over at Toby.

"Can't say that I have." Toby said, realizing that K'Narf was enjoying the fact that it was his turn in the barrel to be out-of-date. Something that was constantly happening to all of them in the extremely fast changing environment they lived and worked in. "You gon'na stop grinning and tell me, or ya gon'na make me ask?"

"Okay, okay. When a male squid is courting a female he constantly changes his coloring as he swims around her. The side of him that would be seen by another male is always his fighting colors, and the side of him that is seen by the female is always his softer "lets be friends and lovers" colors."

"The Lady's new paint job does that?"

"Yes, and more. If you were to see the Lady cross in front of the moon right now, what you would see, if you could spot it, would be a soft blur of movement across the surface of the moon. The Lady's computers would be constantly updating her colors to reflect the specific surface features she was passing over. If you were to look at her against a background of stars, she would appear as that part of the sky based on the exact angle you were looking at her."

"The exact angle?"

"Yes, her right side will always look exactly like what the cameras are seeing on the left side based on whatever angle you may be watching her from." K'Narf said, grinning from ear to ear.

"Sounds like that magic car the kids flew in Harry Potter." Toby said, seeing in his mind what K'Narf was describing.

"What kids? What magic car?"

"It was in a movie, and you don't watch movies so forget it." Toby said, as he deftly changed the subject. "Do you mind if I ask you sort of a personal question?"

"Guess that would depend on the question." K'Narf said, as they cleared the satellite belt and he punched in their destination to the Lady's new computer. Noticing as he leaned back how quickly Toby's mood had changed.

"How do you do it? I mean, how do you understand what's about to happen? I know you see it as prophecy, but how do you see it like you do?" Toby asked, struggling to ask his questions in a way that would let K'Narf see what it was he wanted to know.

"You don't believe in simple questions do you?" K'Narf asked, looking over at Toby, knowing exactly what he was asking.

"I'm sorry, if it's too personal you don't have to..."

"It's not too personal, it's just a bit complicated." K'Narf said, closing his eyes as he put his thoughts in order. "What do you think prophecy is?" He added, keeping his eyes closed as he asked the question.

"Prophecy is where God said something was going to happen before it happens." Toby said, looking over at K'Narf. "That's right isn't it?"

"Yes, it is. Another way of putting it would be that it's history that hasn't happened yet. God, through His prophets, has told us things that are going to be coming upon the world partly to strengthen our faith when they happen as prophesied, and partly so we could prepare for them should we be fortunate enough, or unfortunate enough depending on your point of view, to be living when they take place."

"Yes, I understand that, but I still don't know how you..."

"Reductive logic, let's start with reductive logic." K'Narf said, interrupting Toby as he opened his eyes and looked out into the darkness of space. " A lot of it's just plain simple reductive logic. Let me explain. Most of prophecy cannot be understood except by those who live through, or are about to live through, the actual event. What I mean to say is that until the pages of time and history have turned to the point of the prophesied event actually taking place, it cannot be truly understood because it cannot be logically figured out."

"I'm not sure I..."

"Okay, let me try and explain it this way." K'Narf said, interrupting Toby for the second time. "This is one you should know. There was a prophecy that Israel would become a nation and be birthed in one hour. Now, no nation in history had ever become a nation in one hour before and until it happened it was an inconceivable event. Then six million of your people were slaughtered and the rest turned into refugees. The leaders of the world were forced by conscience to give them a small slice of Palestine to call home, and it happened in the time it took them to vote, one hour. A logical outcome of the events that had taken place, but one that could only be understood by those who lived through the holocaust and the subsequent birth of the Nation of Israel."

"I know my nation's history, but that doesn't explain how you see things that are going to happen?" Toby said, going back to his original question again.

"No, I guess it doesn't, does it?" K'Narf sighed, trying to figure out how to explain his relationship with the Holy Spirit. "When I was six years old my mother gave me over to my grandmother to raise. I was, as T.J. likes to say, mentally challenged at that time. The very first evening I spent with her she sat me down and started reading the Bible to me, two pages every evening for the next twelve years come rain or shine."

"But what's that got to do with..."

"This is difficult enough, just let me talk and see if you can figure it out. Okay?" K'Narf asked, opening his eyes and looking over at Toby. "Good, now where was I, yes my grandmother read two pages of the Bible to me that very first night and it was like your going to the movies. I instantly saw, in pictures and full color, everything she read to me. I accepted the Lord immediately and He accepted me and I've had a strong personal relationship with Him through the Holy Spirit ever since."

"You have a personal relationship with the Holy Spirit?" Toby asked, just a touch of surprise showing in his voice.

"Yes, don't you? I mean when you pray don't you feel His presence?"

"Not sure, I mean I pray, but I couldn't say that there's a close personal relationship involved."

"You don't feel His presence?" K'Narf asked again.

"Not really, what do you call feeling His presence?" Toby asked, realizing that this may be the answer he was looking for.

"Well, for me, it's like wearing a hat when I'm praying. I feel the pressure of the hatband around my head, but of course there's no hat, just a real light pressure as though I was wearing one."

"And that's what you call feeling His presence?"

"Yes, what exactly were you expecting?" K'Narf asked, turning his head so he could look at Toby. The pressure pushing him into his seat was now up to three-point-nine G's where it would stay for the next four hours.

"Not sure, but how does this pressure help you understand prophecy, I mean like you always seem to see what's coming or about to happen?" Toby asked, carefully turning his head so he could return K'Narf's look. "Does the Holy Spirit tell you when something's going to happen?"

"No, at least not directly."

"Not directly? Could you explain that?"

"This gets me back to deductive logic. If I figure something out logically, based on actual events that have happened or are happening in the world, and I take it to the Lord in prayer and ask Him if this is what the prophet meant, I don't get anything even resembling a direct answer; however, if I'm wrong there's always this sense that I'm wrong. I don't know how to explain it, but if I'm wrong I seem to know I'm wrong."

"You know when your wrong, but not when you're right?"

"Right, the feeling I get when I'm right is more like *"if you know your right, why are you asking me?"*

"Can you give me an example of your being right?" Toby asked, realizing he was beginning to understand, even though he knew he really didn't. Irena would be proud of him.

"Yes, like when I figured out that the final beast, which was a lamb with two horns that spoke like a dragon, was the United States."

"The United States?" Toby asked, serious surprise showing in his voice.

"Yes, you know that a beast in prophecy always refers to a country and its associated military, political, economic and religious systems, *not a man* as some people believe. The lamb with two horns described by John obviously represents a Christian nation since it's a lamb, and the two horns are two ruling powers within that nation, which in the case of the United States are the Democrats and the Republicans. Speaking like a dragon is our military since we are the only true super power left in the world.

"How do you figure that?"

"Easy, England, which was the previous beast, remember when the sun never set on the British empire, well because of recent budget cuts and such I doubt if she even has one carrier left in her once mighty Navy. The United States on the other hand has no less than twelve atomic powered super carriers with each one of them surrounded by a full battle group."

"And the Holy Spirit told you this?"

"No, the Holy Spirit just didn't tell me I was wrong?" K'Narf said, beginning to feel slightly exasperated.

"It sounds like me and Irena." Toby said, his voice tinged with a sudden understanding. "Irena always wants me to understand her without her having to tell me. It's kinda like you said, if I guess wrong I can tell by her reactions that I'm wrong, but if I guess right she doesn't react at all. That's the way I know that I'm right." Toby said, trying unsuccessfully to grin under three-point-nine G's of continuous thrust.

"So, your analogy is that the Holy Spirit reacts like Irena, like a woman?" K'Narf asked, unable to force a grin anymore than Toby could.

"I didn't say that!" Toby exclaimed, trying to shake his head in the negative. "And please don't tell that to the Holy Spirit!" He added.

"Why?"

"I'm a Jew, all my life I've been taught to respect, love and fear the Lord. I don't now, and never have had, a relationship like you've described. You talk about the Holy Spirit and the Lord like they were your best and oldest friends. I'll be honest, I think I'm a little envious." Toby said, instantly recognizing the truth of his own words.

"Well, you need to get over your fear and start talking to them as friends, which is exactly what they are, or you may find yourself seriously embarrassed when this whole mess is over with." K'Narf said, letting Toby know by the tone of his voice that he was dead serious.

"What's that supposed to mean?" Toby asked, struggling to turn his head so he could look at K'Narf.

"You do remember that when the final trumpet blows signaling the end, the Lord will send his angels out to pick up any of His that are left alive, you know the ones he calls His remnant, the ones that didn't give obeisance to the beast, before the Father executes his bowl judgments and remakes the world."

"Yeah, I know that, why?" Toby asked, trying to catch K'Narf's new train of thought.

"Then you should know that after the world settles down, the Lord is going to return with all those souls that were under the alter, including the remnant His angels picked up, and they are going to build a new Jerusalem where He is going to live and reign with them for a thousand years."

"Yeah, I know that too." Toby said, obviously not getting what K'Narf was driving at.

"Then you should also know that every year each surviving nation will be required to send a delegation during the seven days of the Feast of Tabernacles and present themselves to Him. You do remember the part in the Bible where every knee shall bow?"

"Yes, but I'm still not following you." Toby said, struggling to get the point K'Narf was trying to make.

"We are a people, a nation of glow heads you might call us, and you are one of the glow head leaders." K'Narf said, spelling it out for Toby in no uncertain terms. "And when the time comes for us to present ourselves to the Lord as representatives of our people, you will be front and center with me, and don't you think it will be more than just a little embarrassing if the Lord walks up to us and has to ask you to introduce yourself."

"You're serious aren't you?" Toby asked, getting the feeling that this was one of those times when he wasn't wrong.

"Yes, but I'm more serious about getting some sleep. I don't want to face Brent with a mind numb from lack of sleep." K'Narf said, as he turned his head to the forward position and closed his eyes, wondering as he did so if he had managed to answer Toby's original question.

"How?"

"How what?" K'Narf asked, refusing to open his eyes.

"How do I talk to the Holy Spirit?"

"Tonight, when your on your knees, start your evening prayers by saying Hi, I'm Toby, mind if I tell you what happened to me today and then tell Him everything, including thanks."

"Thanks, thanks for what?"

"Whatever your thankful for, like yesterday when I went to Wal-Mart's, I asked for a close parking spot and He opened one right up front just as I pulled in, so last night during prayers I thanked Him for it." K'Narf said, as he tried to get a little more comfortable. It was going to be a long four hours and he really did need some sleep.

"Well, you gave me a lot to think about." Toby said, closing his own eyes as he too tried to get comfortable, wondering if his swirling mind would allow him to sleep after what K'Narf had told him, especially the part about introducing himself to the Holy Spirit and saying hi, he thought.

Chapter Two

"K'Narf, wake up, we're here." Toby said, as he opened his eyes and watched the Whale as it grew in front of them. "And there's Brent's red van coming in, we're right behind him. Hey, can he see us?" Toby asked as K'Narf opened his eyes and sat up.

"No, I don't think so, why?" K'Narf asked, stretching.

"See if you can slip in behind him without him noticing us."

"Why?"

"Just do it, okay." Toby said, sitting up straighter and grinning over at K'Narf. "Closer."

"Closer?"

"Yeah, about twenty feet." Toby said, his eyes sparkling as he looked over at K'Narf. "He's Head of Security. Be a bit embarrassing if we could slip through the airlock with him without him noticing us, wouldn't it?"

"Let's make it ten feet." K'Narf said, as he carefully maneuvered the Lady into the shadow of Brent's red van and followed him into the eye of the Whale.

"Careful, he's opening the airlock. He'll have to move up enough to give us room to get in with him." Toby said, getting nervous as the airlock tunnel closed in on them.

"It's a hundred feet long, shouldn't be a problem."

"We're in! You're going to have to move quick so you don't get caught by the exit door when he closes it."

"I know, just hush!" K'Narf said, as the green light came on and the inner airlock door started to raise."

"We're in!" Toby said, and he didn't see us. "Circle around and land facing him. I want to see his face when we de-cloak."

"De-cloak?" K'narf asked, looking over at the smiling Toby.

"That's what the Klingons called it. They cloaked and de-cloaked their war birds all the time." Toby said, laughing at K'Narf's confusion. "Forget it, it's another one of those movie things."

"Cloaking and de-cloaking. I like that." K'Narf said, as he gently set the Lady down on the landing pad with her nose facing the driver's door of Brent's van. "He's getting out, shall I de-cloak her?"

"Do it!" Toby said, as he watched the expression on Brent's face, as a monstrous black van suddenly appeared right in front of him.

"Got him!" Toby said laughing.

"Come on, we need to go talk to our Head of Security about letting an unauthorized vehicle follow him into the Whale without proper clearance." K'Narf said, grinning as he opened his door and stepped out of the Lady.

"Sounds like a reasonable charge to me." Toby said, opening his door and stepping out along with K'Narf. "I think he's as shocked to see us as he was the Lady." He added, looking at the expression on Brent's face.

"How?" Brent asked, leaning back against his van for support.

"Glenn Miller." "K'Narf said, grinning at Brent.

"What about Glenn Miller?" Rudy asked, as he landed literally in front of K'Narf.

"Why, do you know Glenn Miller?" K'Narf asked, reaching out and steadying Rudy as he struggled to get his balance.

"Big band leader during the second world war got killed crossing the English Channel." Rudy said, turning to face K'Narf as he straightened up.

"Wrong Miller. Glenn Miller and his painters of renown, as they call themselves, are the ones who painted up the Lady so she could disappear.

"How'd you do that anyway, make the Lady appear out of nowhere?" Rudy asked, reaching over and touching the Lady as if to make sure she was real. "And what are these bumps all over her?"

"Those are sub-miniature cameras. Each one of them controls one square inch of surface paint. Kinda like a pixel." K'Narf said, turning his attention to Rudy's question.

"Pixel?"

"Yes, but the color of the pixel is controlled by the camera on the opposite side. It gives the impression that your looking through her, hence her disappearing act." K'Narf said, noticing for the first time the odd look on Brent's face.

"Bring me any dirt?" Rudy asked, changing the subject.

"Yes, six fifty pound bags, picked'em up at Wal-Mart yesterday. How's your garden doing?" K'Narf asked, looking down at the brightly lit area that was Rudy's garden.

"Great, over a hundred grow-lights and unlimited power now that we've gone to atomic batteries. Won't be needing vegetables from your brother's farm too much longer, either." Rudy said, giving K'Narf one of his ear-to-ear grins.

"Is the conference room available."

"You got about thirty minutes before the next set of Limo's arrive, then you'll have to clear out. Be forty-one newbies waiting to get in." Rudy said, stepping over and reaching for the Lady's loading bay door.

"Forty-one, what happened to thirteen?" K'Narf asked, knowing the conference room could only hold thirteen at a time.

"Too slow, cut a second deck around the conference room so we could seat three Limos." Rudy said, opening the door and pulling out a large green and white bag and throwing it on his shoulder. "Is he all right?" He asked, nodding at Brent.

"He does look a little pasty faced now that you mention it." Toby said, joining the conversation as they all turned and took a moment to study Brent.

"Okay, let's get him up to the conference room. See if we can find out what all this is about." K'Narf said, as he reached out and took Brent by the arm and literally tossed him at the yellow guy rope that led to the kitchen, forcing him to grab it as he ran into it. "T.J. is going to meet us there." He added, jumping at the yellow rope himself, leaving Toby to follow suit.

"What did you two do to Brent?" T.J. asked, as K'Narf and Toby entered the conference room. T.J. already had Brent seated at the head of the table and was in the process of dragging over a conference chair. Throwing her long black hair over her shoulder to get it out of the way, she sat down and scooted up next to him.

"We ain't done nothing! He just showed up that way." Toby said, grabbing a chair and setting down next to K'Narf who had already sat down.

"Well, something's wrong with him." T.J. said, taking one of Brent's hands and stroking it like you would a small child. "What's wrong Brent, what happened to you?" She asked gently as they all looked at the stricken expression on Brent's face as he raised his head and looked directly into K'Narf's eyes.

"I found them." He said quietly.

"Start at the beginning, we'll listen." K'Narf said gently, beginning to realize that Brent was in a mild state of shock.

"I sent out teams to all the monasteries around the world just like you asked me to, then I personally headed for the Sinai and St. Catherine's which is the oldest Christian monastery in the world. Like you said, it seemed to be the obvious choice if John and his friends were hiding out, but getting into their inner sanctums turned out to be a little harder than I thought." Brent said, his eyes coming alive as he started talking.

"Go on." K'Narf said softly.

"I ended up getting a job with an old man called Jakli who delivered supplies to them.

Worked for him over two months before they began to get used to me enough to let me inside when we delivered fresh produce. Then two days ago I arranged for Jakli to have a nerve problem in his back so he had to depend on me to deliver a load of fresh citrus. I was wearing my Kirlian contacts and spotted him immediately."

"Spotted him?" Toby asked, getting an immediate shushing look from K'Narf.

"Yes, he was sitting at one of the long eating benches all by himself peeling one of the red navels I had just brought in. His head was glowing like crazy. I turned off the contacts and walked over and sat down opposite him and said good morning in the scroll language just like you suggested. It worked!" Brent said, giving K'Narf a look that said he was coming out of shock.

"Go on." K'Narf said, catching the surprised look on T.J.'s face.

"He... he froze me!" Brent said, his voice trembling.

"Froze you?" Toby exclaimed, ignoring K'Narf's shush look.

"Yes, he took control of my mind and turned my whole body into what felt like a piece of ice. I couldn't breath or even blink my eyes. He looked at me for all of a full minute then did something to my motor nerves that allowed me to start breathing again."

"Go on." K'Narf said, giving Toby another shush look.

"Then his two buddies showed up. They all sat there with him eating that orange and looking at me. It was obvious that they were talking about me, but not verbally. I figured they were flashing each other. The older one noticed that I wasn't blinking and did something that seemed to very carefully release my eyelids so I could blink, but that was all he released. Then they went after my mind. It felt like my whole life was on rewind. They took me all the way back to my being born; memories that I didn't even know I had." Brent said, shaking his head at the memory of what had been done to him.

"Then what happened?" T.J. asked, caught up now in Brent's story.

"They spent the next two hours walking up through my memories, it was like watching a movie of my life with them as spectators. They especially got a kick out of my first day in school when I took a rubber band and tied that little fat know-it-all girl's pigtail to a bolt sticking out of the back of her desk. When she stood up to answer a question for the teacher her whole desk went over dumping everything, including her, on the floor. I got suspended and my father made me swamp out the stalls in the barn for the whole week I was out of school. The wheelbarrow was bigger than I was and what made it worse was that he kept thanking me for getting expelled. They all thought that was real funny!" Brent exclaimed, showing more and more emotion as he talked.

"What happened to the little girl?" T.J. asked, looking concerned.

"Parents took her out of our little reservation school and sent her to some kind of fancy all girls school and I never saw her again, why?" Brent asked, turning to look at T.J.

"Go on." K'Narf said, giving T.J. a shush look this time. "What did they do next?"

"When they finally got up to how I had found them, they started playing with my Kirlian contacts."

"They took them out?" Toby asked, surprised.

"No, they turned them on and took turns looking through my eyes at themselves. They seemed surprised that they glowed like they did. Then they did something that instantly put me to sleep and when I woke up they were gone and I got out of there as fast as I could and came here." Brent said, seeming to shudder at the memory.

"Then they know all about us." K'Narf said, a worried look on his face.

"Whatever I know, they know. It's as simple as that." Brent said, holding his head in his hands.

"But they didn't kill you." T.J. said, stating the obvious. "And they could have." She added, seeing the meaning of her words dawn on K'Narf.

"But Brent." T.J. said, suddenly reaching out and taking Brent's hand and turning him slightly so he had to look at her. "I'm worried about that little girl you told me about, is that all you remember of her?"

"The little fat girl! Why are you worried about her? She was just a little know-it-all, and she's lucky her parents could afford to put her in a private school where I couldn't get back at her for that week I spent swamping out my dad's barn!" Brent said, raising his voice as he responded to T.J. question.

"Fat girl!" A voice from behind him screamed, causing all of them to turn and look at the girl standing behind them. Raven black hair with a red feather braided into it, beaded headband, white doeskin tunic and trousers with long flowing fringes. A stunning figure with the soft light of the conference room shining upon her; a stunningly angry figure with the soft light of the conference room shining upon her to be more exact.

"T.J., what are you up to?" K'Narf flashed, as he turned and caught a look of sheer glee on T.J.'s face. "Who is she?" He flashed again,

"She's the little fat girl. Her name's Redwing and she's the one I brought up to help Rudy and Wanda."

"You set this up!"

"No, but my grandfather used to tell me that when opportunity knocks, open the door wide and let it in!" She flashed, openly laughing now.

"Fat girl!" Redwing screamed again. "You miserable offspring of a Jackass, I spent eight years in an all white girls school where I was the only Nez Perce and you know what they called me? No! You don't need to know what they called me! You, who can walk upright under the belly of the great snake! No! No! No! You are not worth this feeling of anger. How can one be angry with a thought, with the wind, with nothing?" Redwing asked, as she abruptly turned and left the room. All eyes turned to look at Brent.

"I think he's in shock again." Toby said, looking at the stunned expression on Brent's face.

"Sorry to break this up, but the Limos are here and we need to adjourn to the kitchen." T.J. said, standing up and pushing her chair back to where it belonged. "The newbies are going to need the conference room." She added as she motioned for K'Narf to help the still stunned Brent.

"Give me a hand here." K'Narf said, Taking Brent by his right arm and standing him up. "You get his left and we'll carry him out." He added, grinning at Toby.

"I can walk!" Brent said, suddenly coming to life. "Let go!" He added, pulling his arm out of K'Narf's grip and heading for the door.

"Feisty little fellow, isn't he?" Toby said, grinning as he followed K'Narf out.

"Rudy!" K'Narf exclaimed, as Rudy came flying into the kitchen just as he stepped out of the conference room. "We're moving over to the picnic table. Why don't you join us?"

"Okay, just let me get washed up a little." Rudy said, landing catlike and then seeming to glide across the floor as he headed for his bedroom.

"Where did T.J. go?" K'Narf asked, as he sat down where he would have a good view of the incoming newbies.

"She and Wanda are in with Redwing. Seems our friend here really upset her with his "fat girl" remark." Toby said, laughing at Brent.

"I was set up!"

"Told you, you shouldn't have used T.J.'s picture for show and tell." Toby said, grinning.

"Hey, here they come." K'Narf said, as the newbies started dropping into the kitchen, most of them bouncing off the floor and floating in the air until one of their buddies grabbed them and pulled them down.

"K'Narf, notice how they barely give us a glance, that's because they have no idea who we are, but when they come out of orientation they'll be all gaga when they recognize us, especially you."

"How do you know that?" K'Narf asked, realizing that Toby was deliberately sitting so his back was to the incoming newbies.

"Happened to me last week. It was downright embarrassing."

"T.J.," K'Narf said, as T.J. came out of the second bedroom. "Is she all right?" He asked, as he stood up and let T.J. slide in next to him.

"She's fine, it was just a shock when she realized that Brent was the little boy that caused her to be sent to that all girls school." T.J. said, smiling at Brent as she put her hands together and gave him a small bow of recognition.

"You set me up, you did that on purpose didn't you?" Brent said, as he returned her bow with a slight nod of his head.

"The Yin and Yang are in harmony, my inner world is now at peace." T.J. said, which was her way of saying yes.

"Are you two going to be friends again now?" K'Narf asked, looking at each of them in turn.

"We always have been." T.J. said, glancing up at K'Narf as if he had said something foolish, then turned to look at Brent again. "You will need to get with Minh Le and figure out how to protect your mind and block that kind of mental attack." She added, suddenly all business.

"I know, if they hit our fighters with a mental attack like that it wouldn't be a fight, it would be a slaughter with us on the receiving end." Brent said, grimacing at the thought. "I think I ought to be going. This is too important to wait on." He added, as he started to scoot out.

"No!" K'Narf suddenly exclaimed, not realizing that he had made everyone jump with his sudden exclamation. "I'd like you to hang around until the newbies come out. There's one of them I want to talk to, and I'd like you to be here when I do."

"Why?" T.J. asked, stating the question for all of them.

"Premo just flashed me some information about one of them and I want to talk to her." K'Narf said, not knowing exactly why himself.

"Her?" T.J. asked, the female pronoun catching her attention.

"Yes, she should be easy to spot, Premo says she's the only one that still has grey hair and wrinkles." K'narf said, adding a sense of mystery to his sudden decision.

"Rudy, have a seat." Toby said, sliding over to give Rudy room to slide in. "You're looking fantastic. Bonded life must agree with you." He added grinning.

"I didn't think it was possible to be this happy and live." Rudy said grinning. "Oh good, Redwing's gettin' coffee." He added, as everyone turned and realized that Redwing had come out of her bedroom with Wanda. Dressed now in form fitting sweats, the standard dress for the Whale, she was holding a tray and Wanda was loading it up with coffee mugs, soy cream, sugar and a coffee pot. They all watched as Redwing calmly took one of the mugs off the tray and carefully put it back on the shelf, then turned and started walking toward them.

"Thank you Redwing." Rudy said, as she handed him a mug and very carefully filled it with steaming black coffee, exactly the way she knew he liked it. Then, smiling at each of them in turn as she carefully walked around the table and handed each of them a coffee mug and filled it with coffee, letting them add cream and sugar per their own taste, then walked away without saying a word.

"Am I missing something here?" Rudy asked, looking at Brent who was the only one sitting at the table without a mug of coffee.

"He called Redwing a little fat girl!" Toby said, grinning as he very deliberately rubbed salt into an open wound. "Just trying to get my Yin and Yang back into balance." He added, grinning at Brent.

"A man would have to be blind, dumb and stupid to call that girl fat." Rudy said, glancing at the retreating form of Redwing and then looking over at Brent.

"It's not like it sounds, I was set up!" Brent said, glaring at Toby and then at T.J.

"You're gonna end up paying for that." T.J. flashed, shaking her head at Toby.

"I know, but right now it's worth it." Toby flashed back.

"Here they come." K'Narf said, as the first of the newbies came out of the conference room, immediately stopping and staring as she realized who she was looking at. The newbie coming out behind her was instantly pushed by the girl coming out behind him, who in turn stopped and stared as she too realized who she was looking at. This process continued as each of them emerged from the conference room until finally the one K'Narf had been waiting for came out.

Like Toby had said, their stares were downright embarrassing.

Chapter Three

"That's her." K'Narf said, nodding his head at a tall stately woman with long grey braids coiled on top of her head who had just emerged from the conference room. "Rudy, could you go ask her to join us, I'd go but I'm afraid I'd start a ruckus."

"Not a problem." Rudy said, as he slid out of his seat and did a short quick soft shoe shuffle over to the line of newbies. They all watched as he explained to the woman K'Narf had pointed out that he, K'Narf, wanted to talk to her. You could tell that Rudy had used his name by the reaction of the other newbies. Then they both turned and Rudy brought her over to the table. "This here's DeeDee. Scoot over Toby, and let the lady set down." He added, as he took her arm and seated her next to Toby, then did a little two-step down to the other end of the bench and slid back into his own seat and took a sip of his still very hot coffee.

"DeeDee." K'Narf said, breaking the ice. "Is that a nickname?"

"Yes sir. Full name's Dorothy Dean Alexander. Father wanted a boy so I got Dean for a middle name and I've been called DeeDee since I can remember." DeeDee said, very obviously flustered.

"Do you mind talking with us for a few minutes?" K'Narf asked, trying to put her at ease.

"No sir, as long as I don't miss my ride."

"Not to worry, I've already flashed your driver and asked her to wait for you." K'Narf said, stopping his coffee mug just short of his mouth. "They're all heading for the chapel and that will keep them busy for a little while anyway." He added, continuing the movement of his coffee mug to his mouth as he took a sip.

"Yes sir." DeeDee said, looking over at T.J. and Brent.

"I'm sorry DeeDee, this is my wife T.J., Mister Brent Conrad, Rudy Marcus you've already met and your sitting next to Toby Zebulon."

"Yes sir, I recognize them from orientation."

"Please, call me K'Narf." K'Narf said, feeling uncomfortable being called sir.

"Yes sir, uh, Mister K'Narf. What is it you wanted to talk to me about?"

"I was curious about why you were so angry at Brent's pick-up team when they found you and brought you in?" K'Narf asked, getting right to the point.

"I wasn't angry with them, I was angry at God, and still am!" DeeDee said, no longer the timid little old lady as she slapped the picnic table with her hand making everyone jump.

"You're angry at God?" T.J. asked, noticing how DeeDee's dark green eyes seemed to be sparkling with small yellow flecks as she spoke. She has hazel eyes she thought, flashing K'Narf to look at the color of her eyes.

"He put me through the proverbial wringer for no good reason, and until He explains why I'm gon'na stay mad at Him." DeeDee said, pointing up as though she was pointing at God. "Oh, don't look so shocked, He's used to me being mad at Him." She added, calming down as quickly as she'd gotten riled up.

"What do you mean He's used to it?" Brent asked, his curiosity getting the better of him.

"I'm sorry, shouldn't have said what I said." DeeDee said, trying to get out of answering Brent's question.

"No, please, I'd like to know too." K'narf said, slowly taking another sip of coffee. Trying in his own way to put DeeDee at ease.

"Okay, but remember, you handed me the podium." DeeDee said, taking a deep breath.

"I'll hit the buzzer if you go past your time." Toby said, chiming in as he shifted sideways so he could get a better look at DeeDee when she talked.

"When I graduated nursing school back in fifty-three." DeeDee said, taking a deep breath. "I took a job working the midnight to eight in the morning shift in the emergency room at the old Trinity Hospital. I worked that shift for the next forty-two years until I retired in ninety-five on my sixty-fifth birthday. I won't even try to tell you what came through those doors during those years, but I will tell you about one of the times I got especially mad at God. Okay?"

"Go on." K'Narf said, nodding at her.

"One night, just after my shift started a woman came in holding a baby. He was the prettiest little boy you could imagine. Bright blue eyes and blond hair and he was gagging and turning blue. She handed him to me and told me his drunken father had decided his new son should have a drink with him and poured a whole shot of one hundred and eighty proof vodka down its little throat. I didn't do it, but I'm the one God gave that little baby to when it gasped its last breath and died in my arms. I was so mad at Him, I don't think I talked to Him for at least a week, maybe two." DeeDee said, tears welling up in her eyes at the memory.

"Oh sweet Lord Jesus." T.J. said, covering her face with her hands. Trying not to see the image that DeeDee's story had placed in her mind.

"So why are you mad at Him now?" K'Narf asked, pulling everyone back to the situation at hand.

"Like I said, He put me through the wringer for no good reason that I can see, and I'm gon'na stay mad at Him until He tells me why." DeeDee said, quickly forgetting about the long dead baby.

"What do you mean He put you through the wringer?" K'Narf asked, knowing that DeeDee was comparing what God had done to her to an old method of putting laundry through a wringer to squeeze the water out. His grandmother had had an old washing machine just like that when he had first gone to live with her.

"Longer story, sure you want to hear it?" DeeDee asked, suddenly looking unsure of herself.

"Yes, please, I'd very much like to hear it." K'Narf said, visibly relaxing as he took another sip of coffee, still trying to put DeeDee at ease.

"Well, after I retired I joined a small group of seniors at church in a weekly bible study aimed specifically at prophecy and the end times that were coming." DeeDee said, seeing K'Narf's eyes light up at the mention of prophecy. "There were seventeen of us, eight couples and myself. I was always the odd one out so I was used to it."

"You never got married?" T.J. asked, remembering the many years she had spent alone.

"Tried it once. Married an accountant; said he loved me but it didn't last out the year."

"What happened?"

"I worked nights and he worked days. I'd come in dead tired, usually with blood on my scrubs, just as he was all dressed up pretty for work. Then he'd come home dead tired just as I was getting up and feeling good. I wouldn't change shifts and he wouldn't take a night job. Just didn't work out, that's all." DeeDee said, surprised to see empathy for her in T.J.'s eyes.

"Please, go on." K'Narf said, flashing T.J. a shush sign which got him a silent dig in the ribs.

"Well, when we figured out that the three sixes were the three men mentioned in Daniel eleven and the fact that the United States had to be the lamb with two horns and that the Democrats and Republicans have to be the two horns."

"How did you figure that out?" Toby asked, surprised by DeeDee's comments since he had just heard the same thing from K'narf on the trip up to the Whale.

"Toby!" K'Narf scolded, giving Toby a "shush up" look.

"No, it's all right. Duane, our group leader, called it deductive logic, I called it plain old horse sense where you put two and two together and come up with four and that's when Judy came in with that book 'The China Study' and we all became diehard Vegans and what the world today calls preppies."

"Vegans and preppies?" This time it was Brent that reacted to DeeDee's words.

"Yes, we knew by this time that we were getting close to the end, and the coming back of the Lord and we didn't want to die and miss it so we all became Vegans. Figured if we could add ten or fifteen years to our lives we would have a good chance to see His return. Even hired a chef to come in and teach us how to cook Vegan. It was a lot of fun."

"Would you like something to drink?" Redwing asked, addressing DeeDee. She had walked up on them while they were talking with a pot of steaming coffee in her hand.

"I would love a cup of herbal tea dear, if you have any." DeeDee said, smiling up at Redwing.

"I'll bring it right over." Redwing said, as she walked around the table carefully topping off each coffee mug.

"I don't know what you did to that young girl Mister Conrad, but I'd be nervous if I was you." DeeDee said, having caught the look Redwing had thrown at Brent.

"He called her..."

"Thank you Toby, I was looking for a volunteer." Brent said, glaring at Toby as he cut him off.

"Volunteer?"

"Yeah, someone I can use as a victim when Minh Le and I demonstrate how to protect our minds from being frozen in an attack. I really appreciate your volunteering for our demonstrations." Brent said, grinning now that he had figured out how to get his own Yin and Yang back into harmony, at least where it concerned Toby.

"But I didn't..."

"Ah, but you did, I told you to be careful." T.J. said, flashing Toby a picture of himself stretched out on a torture rack.

"Okay, okay. Sorry DeeDee, they get a little carried away sometimes. Please go on." K'Narf said, smiling at DeeDee and frowning at Brent and Toby as he tried to get her back into her story. "What did you mean by becoming a preppie?"

"That means someone who squirrels away food in case of a major disaster." Toby said, speaking up as he remembered what he had read on the Internet about preppies.

"And why did this make you angry with God?" T.J. asked, still trying to understand why DeeDee had said she was mad at God.

"Oh, I'm not angry at God about any of that stuff." DeeDee said, smiling at T.J. "I'm angry with Him because He had me break up with all my friends, sell my house and move to New York for nothing!" DeeDee said, anger again showing in her voice.

"Why in the world would you do that?" T.J. asked, surprised by DeeDee's outburst.

"God said in His word that when the abomination is in the temple we were to '*enter thou into thy chambers, and shut thy doors about thee; hide thyself as it were for a little moment, until the indignation be overpast*'. Well, Daniel says that '*indignation*' is going to be in that temple for one thousand, two hundred and ninety days. That's a little over three-and-a-half years that the antichrist and his beastly systems are going to be hunting us!"

DeeDee said, her voice sounding almost like a growl. "And we realized that if one of us got caught they would make them tell where the others were hiding, so we split up. I decided to go to New York where they have a reputation of ignoring their neighbors. Sold my house and everything." She added, shaking her head as she remembered the beautiful little house she'd lived in most of her adult life.

"Hunting us?" Brent asked, picking up on the image of someone hunting him.

"Yes, hunting us, just like Nero did back in the days of the Roman Empire. You know Nero wasn't crazy. Him and his senators were actually pretty sharp. Old Genghis Khan had been sneaking his men into the Roman Empire as everything from peddlers to soldiers in the Roman Legions. Then they would sneak out and group up at night to raid and pillage outlying Roman estates, then be back in their beds by daybreak. They were destroying the Romans from the inside. Well, old Nero and his cronies figured out that Genghis' men believed in one god and one god only. That's why they came up with that grain offering to Nero gimmick, they knew Genghis Khan's men wouldn't do it, what they didn't realize was that neither would the Christians."

"You're saying the Nero wasn't really after Christians." K'Narf said, nodding his head in agreement as he realized she was right.

"Not when he started, one of Nero's Governors even wrote a letter to him describing how he had rounded up a bunch of fundamentalists he believed to be Genghis Khan's men along with a group of people calling themselves Christians. He told Nero he wasn't sure what to do with them so he just had them killed along with the others he'd rounded up. Nero wrote back that he wasn't sure either, but it sounded like the right thing to do. It's all in the history books."

"And you think that's going to happen again?" Brent asked, glancing over at K'Narf.

"Just as sure as we're sitting here talking!" DeeDee said, slapping the table for emphasis again. "Only this time they are going to be hunting suicidal terrorists instead of Genghis Khan's men, and just like last time we're going to get caught in the middle. It's already started." She added, noticing that K'Narf was agreeing with her.

"Yes, I read where some of the various religions have already started calling Christian fundamentalists terrorists." K'Narf said, remembering the article he'd read about the World Council of Churches.

"That's right, and when the time comes, they won't differentiate between a suicidal terrorist and a Christian fundamentalist any more than Rome did. If they catch you they will kill you, it's as simple as that."

"Who exactly are they?" Brent asked, still trying to figure out who would be hunting him.

"They are the *system*!" DeeDee said, saying the word '*system*' like it tasted bad. "First they'll barcode everyone with a chip in the back of their hand or in their forehead, and when that doesn't work they'll have to come up with some religious something or other that will force the terrorist to reveal themselves as fundamentalists who believe that their god is the only way. Just like Nero did with his little grain offering, problem is they're going to catch us in the same net."

"Why does the chip have to be in the back of the hand or in the forehead?" T.J. asked, stating a question she'd long held about the so-called mark of the beast.

"Simple enough. Those are the only two places that have enough variation in temperature to keep a lithium battery charged up, and the chip they plan to implant has exactly that kind of battery. Think about it, how do you check a child's temperature; you place the back of your hand on the forehead, not by touching the back of your hand to their little fannies."

"Why does the battery have to be lithium, can't it be something else?" Brent asked, trying to imagine a different battery.

"No, you need one that is self charging and besides God says that his first bowl judgment is to give all of the antichrist's followers a grievous sore, not sores, just a single grievous sore."

"I thought that prophecy meant boils." Toby said, remembering a sermon to that effect.

"Oh, give me a massive break! The first bowl judgment dished out by a very angry God is going to be a bunch of boils that could be lanced and sterilized by any first year med student! No! He's going to pop all of those lithium batteries in a single stroke and destroy the antichrist's control of both his military and all the people under his rule. A grievous sore, one that won't heal the Bible says, and a dose of lithium under the skin would cause exactly that kind of sore! God is going to destroy the antichrist's control by destroying all of his implanted chips, it's that simple!" DeeDee exclaimed, turning to look at the hapless Toby as if to say "how stupid can you get".

"You said, barcode?" Toby asked, trying to change the subject, noticing as he did that K'Narf seemed to be agreeing with everything DeeDee was saying.

"Yes, nine numbers for your social security number, and three control bar numbers. They're going to barcode everyone with their social security numbers. No more illegal anything, much less illegal immigrants." DeeDee said, visibly calming down.

"I understand and I agree with everything you've been telling us DeeDee." K'Narf said, as DeeDee turned away from Toby and faced him. "But could you tell me exactly what it was you and your friends decided to do, and how you were going to do it?" He asked, knowing there was something specific the Holy Spirit wanted him to find out.

"Yes sir, uh, I mean Mister K'Narf." DeeDee said, stammering a little; giving the impression that she was embarrassed over her outburst about the boils. "Like I said, we split up so nobody could know where anyone else was hiding. I sold my house, changed my bank account to the Chase Bank in Manhattan, and went down and stopped my social security."

"Why did you do that?" T.J. asked, surprised that anyone would voluntarily give up their social security money.

"Didn't need the money, and I figured it was just another way they could track me through the system." DeeDee said, pausing to see if there were any more questions. "Took me about three months before I found a new seniors only apartment building that was still under construction. I bought a two-bedroom unit on the third floor facing south; high enough up to feel safe, but low enough that a ladder truck could reach me if there was a fire. Had them enclose the balcony and turn it into a miniature greenhouse and moved in."

"Go on." K'Narf said, realizing that this was what he had been waiting for.

"Not much left to tell really. I turned the second bedroom into a storage room with real nice stainless wire storage racks and started stocking it. We'd all made up lists of what and how much we would need in order to hide out for Daniel's one thousand, two hundred and ninety days, so shopping for the stuff wasn't hard to do. Then it was just a matter of waiting until it was time to lock the door and stay inside like the prophet said *'enter thou into thy chambers, and shut thy doors about thee'*." DeeDee said, looking at K'Narf as if to say is there anything else you want to know.

"Could you describe your list of stuff for me?" K'Narf asked, seeing the puzzled look on Toby's face.

"Well, yes, I guess I can. We had each worked on our list for a couple of years so it included just about everything you can think of from toilet paper to lipstick. Being a Vegan I had to store a lot of raw grains like wheat so I could grind my own flour, oat groats for oatmeal and dried raisins to go with it. Lots of nuts and legumes, pinto beans, rice and stuff like that. Lots of dried stuff too, potato flakes, pasta and all the various spices that I like to cook with. I had a thyroid problem so I had to have enough pills to last and then I had my treadmill for exercise. Bought a bunch of those trail DVD's so every time I exercised I could walk down a different trail. Of course everything was on a FIFO system so nothing would go stale. Oh, and cat food and litter for Sammy"

"FIFO system?" Brent asked, surprised by the term.

"Yeah, you know first in first out, that way you always use up the oldest stuff first. Had all the stuff for my greenhouse garden too, you know, seeds, fertilizer, potting soil and stuff like that. After that it was just a matter of waiting it out."

"Waiting it out?" T.J. asked

"Yeah, I loved going to off Broadway plays, museums, central park and sometime a movie. Couldn't eat out though, being a Vegan I pretty much had to cook my own food, and even after it started and the two witnesses showed up, it would have still been three-and-a-half years before I'd have had to do a final stock up and go in and lock my door." DeeDee said, looking askance at K'Narf to see if he agreed.

"I'm still not sure who it is that is going to be hunting you." Brent said, still working on the person or persons that would be doing the hunting.

"It's not a person Mister Conrad, it's the system itself that will be doing the hunting as you put it. There won't be a boogeyman or some kind of storm trooper chasing you down the street shooting at you. Let me explain. Today, when you walk into one of the large box stores, they read any chips that may be imbedded in your clothing or in anything else you may have on your person. If the chip is one that they implanted in something you're wearing that you bought from them, they instantly know who you are and can track you as you walk through their store. That way when you check out, they can target the coupons printed on their sales slips, or the ones they mail or e-mail to you, based on what areas of the store you showed an interest in, and a scanner can cover an area of slightly over three hundred thousand square feet so the whole store is covered."

"But how does that..."

"Let me finish. When the day comes that we're all bar-coded, all they have to do is put scanners in all the other stores that don't already have them, including gas stations and convenience stores. That way no matter where you go you're going to get scanned. Now, after they set-up whatever it is they plan to set-up in the new Jewish Temple, and demand that everyone gives obeisance to it, your giving of obeisance will be recorded in your personal computer profile showing you did it, just like the piece of paper the Roman citizen had to carry showing he had made a grain offering to Nero."

"And if I refuse." Brent said, beginning to get the picture of what was going to happen.

"If you get scanned, the computer instantly checks your profile and an alarm is put out to the police that you are a possible suicide bomber and they will treat you that way."

"Suicide bomber?"

"Yes, as far as the system is concerned, if you haven't given obeisance to this new god they've set-up, it means you're a fundamental terrorist and may be wearing a suicide vest and you can scream out that you're a Christian all you want, it won't make any difference."

"Then what happens?"

"If you're not killed on the spot, and you still refuse to recant and recognize their new deity, you'll be processed out." DeeDee said, looking grim.

"Processed out?" This time it was K'Narf asking the question.

"It's real simple Mister K'Narf. They'll just delete your computer file, picture, fingerprints, driver's license, bank accounts, medical file, military history if you had one, social security number, birth certificate, literally everything that had been accumulated over the years and put into your personal profile. Electronically you will cease to exist, then they will dispose of you physically using what they're calling the green option."

"Green option?" Again it was K'Narf asking the question.

"Yes sir, one of the options morticians are offering today is what they call the green option. You have traditional burial with a coffin; cremation with its high cost of fuel and left over ashes and now the green option. The green option is simply a steel chamber where a body can be placed, then the chamber is filled with a potassium hydroxide solution, heated to two-hundred degrees under pressure for three hours and then, with the body totally dissolved, it can be flushed into the sewer system. They call it the green option because there is no high consumption of natural gas, no ashes or fumes contaminating the air and it's cheap. The end result is that the so called terrorist, Christian or otherwise, no longer exists, electronically or physically."

"Thank you DeeDee, now I understand what's going to happen during the time that the Bible says we won't be able to buy and sell." Brent said, giving DeeDee a nod of respect.

"Mister K'Narf?" DeeDee asked, looking at K'Narf to see if he wanted anything else.

"Yes, I think I have an answer for you now." K'narf said, smiling at DeeDee. "You wanted to know why God put you through the wringer as you put it, didn't you?"

"Yes!" DeeDee said, her eyes lighting up along with everyone else.

"What are you up to K'narf?" T.J. flashed.

"Just back me up!" K'Narf flashed back.

"DeeDee, I'd like to offer you a staff position as Head of Procurement. In that position you will be responsible for setting up and filling a warehouse with everything from toilet paper to lipstick, as you put it, for all of us rather than just yourself this time. I believe God put you through what He did in order to prepare you for the job I'm offering you. Will you accept?"

"How, how many people?" DeeDee stammered.

"I'm still working on the one hundred and forty four thousand number, but I'd like you to plan on two hundred thousand and ten years rather than the three and a half you initially planned for. Your warehouse will have to be constructed out here where it will be safe, rather than down on Earth though." K'Narf said, trying to give DeeDee time to assimilate what he had said.

"Out here?" DeeDee said, still trying to grasp what K'Narf was offering her.

"Yes, as a matter of fact, Mister Conrad is in the process of constructing a place similar to what you're going to need for himself. Is that correct, Mister Conrad?"

"Yes I am, and I would be very happy to show my little asteroid estate to her if she would like to hitch a ride back with me." Brent said, offering DeeDee a ride back to earth in his van as well as a short sightseeing trip to his own private asteroid.

"I wouldn't mind seeing that myself." Toby said, in essence inviting himself along.

"Why don't all of you come?" Brent asked. "My asteroid is not that far away, so it wouldn't be that much out of the way."

"DeeDee?" K'Narf asked, as all eyes turned to DeeDee.

"Yes, oh my God yes!" She said, tears welling up in her eyes.

"K'Narf, I just told the limo driver she doesn't have to wait, but DeeDee, before we head out I have one last question for you. It's obvious you've repaired your body, but why didn't you get rid of the grey hair and wrinkles?" T.J. asked, stating the question that had been on her mind since the first moment she had laid eyes on DeeDee.

"That's his fault!" DeeDee said, wiping at her eyes as she pointed at Brent.

"What is this, pick on Brent day?" Brent asked, surprised at the accusation.

"He won't let anyone be part of a fighter team for one of his precious Teardrop Fighters unless they're a bonded couple. The whole place down there is like some kind of weird love boat with everyone looking for someone to bond with, leaving myself looking like this was my only defense." DeeDee said, frowning at Brent, and then laughing. "But I won't be looking like this much longer. I used to have gorgeous red hair. Everyone use to say I looked like Maureen O'Hara." She added, grinning.

"What's a Maureen O'Hara?" K'Narf asked, looking over at Toby.

"It's another one of those movie things." Toby said, grinning.

"Well, we need to get going, so I suggest we go get a third chair bolted into the Lady so you'll have a place to sit." K'Narf said, sliding out and standing up and stretching.

"Mister K'narf." DeeDee said, sliding out and standing up too. "Would it be all right if I spent a few minutes in your chapel. I have some apologizing to do and it's not something I want to put off."

"Please, it's just K'narf, and take all the time you need." K'narf said, understanding her need to talk with the Lord.

"I'll take her." T.J. said. 'We'll meet you at the landing pad." She added, sliding out and taking DeeDee by the hand.

"You know you've created a monster." Toby said, watching as T.J. leaped out of the kitchen followed by DeeDee.

"How so?"

"That woman's now on a mission for God Himself, and God help anyone who gets in her way." Toby said, flashing K'Narf and Brent a caricature of DeeDee on a steamroller with everyone scattering to get out of her way.

"Yes, she will be formidable." K'Narf agreed. Now let's go find a spare seat and get it bolted in so you won't have to squat on your haunches all the way back."

"Relegating me to the back seat are you?"

"No, but you're the one that will have to tell T.J. she has to take the back seat." K'narf said, giving Toby an "I dare you" look.

"What, and risk getting her Yin and Yang out of balance again, not even if you ordered me to." Toby said, as he followed K'narf out of the kitchen.

Chapter Four

"I thought he said it was close by." Toby said, looking around T.J.'s seat at Brent's red van flying in front of them. Noting the fact that they had been moving slowly through the asteroid belt for more than twenty minutes now.

"Remember, it's Brent you're talking about." K'Narf said, catching Toby's eye in the rearview mirror.

"Stop picking on Brent." T.J. said, scolding both of them.

"Oh, so now you're going to protect him." Toby said, grinning. "That's what I call a change of heart, and just because you got even with him doesn't mean we have." He added, poking T.J.'s seat with his new foot, still finding it amazing that he had been able to grow it back.

"Did I tell you that I promised DeeDee one of the kittens?" T.J. asked, ignoring Toby as she looked over at K'Narf.

"Kittens, what kittens, I thought we gave them all away?"

"Judy had another litter three days ago, two girls and one boy. I promised the boy to DeeDee." T.J. said, repeating her promise to DeeDee.

"So that's why Thomas didn't come to see me." K'Narf said, remembering being curious as to Thomas' whereabouts.

"She was still feeling bad about losing Sammy, that's why I promised her one." T.J. said, explaining why she had promised the kitten.

"Who's Sammy?" Toby and K'Narf both asked at the same time.

"Her cat, she told you about him. She'd had him eighteen years when he suddenly passed away. She had him buried in one of those pet cemeteries. She was coming out of the cemetery, after sitting on what she called a sobbing bench all day, when Brent's pick-up teams spotted her." T.J. said, continuing her explanation as to why she had given DeeDee the kitten.

"What made you want to talk to her in the first place?" Toby asked, changing the subject.

"Premo flashed me about getting updated. Said there'd been a lot of information added to his readers and wanted T.J. and me to drop by." K'Narf said, noticing that Brent seemed to be heading toward a monstrous off-white asteroid. "When I told Premo where I was he told me to watch for her. Said she was right out of the cracker barrel." K'Narf said, remembering Premo's flash.

"Cracker barrel?" Toby asked, noticing the large off-white asteroid they were now obviously headed for.

"Yeah, he told me how she was angry as all get out when she was picked up, and had refused to fix up her appearance even though she had already processed the reader that told her how. Felt like the Holy Spirit wanted me to talk to her the second I got Premo's flash about her, so I did."

"Wait until he finds out you put her on staff." Toby said, grinning at K'Narf in the rearview mirror. "Forty-two years running an emergency room. Bet she's seen more blood and guts than Patton."

"Who's Patton?"

"World War Two General everyone called blood-and-guts, course his men said it was his guts and their blood. She just reminds me a little of him. Tough as nails he was." Toby said, remembering the General he had written a paper on when he was studying to become an Israeli officer.

"What's with you two? Always have to be picking on someone. Now leave DeeDee alone. Did I tell you I gave her a Debit card on the general fund?" T.J. asked, seeing Brent's red van disappear into a large crevice in the side of the slowly rotating asteroid looming in front of them. It appeared to be rotating at about the same speed as the Whale.

"Did you tell her that money was '*not*' an object of concern?" K'Narf asked in return, turning to grin at T.J. as he asked the question.

"No, I told her that there would be funds available for whatever she would need." T.J. said, giving K'Narf a snapish look. "I will not find it necessary to worry about her spending money unnecessarily."

"Careful, you're going to get her Yin and Yang out of balance again." Toby said, as they followed Brent into the blackness of the crevice.

"It's harmony, not balance." T.J. said, correcting Toby as K'Narf turned on their headlights and followed Brent as he landed on a ledge in order to match the spin of the asteroid, following the same procedure needed to enter the whale. "And I will enjoy watching Brent use you for an example in practice." She added, reminding Toby of his soon to be fate.

"Would you look at that!" K'Narf said, as Brent lifted and flew into a perfectly cut circular entrance tunnel. "Wow!" He added, as they approached the first airlock door and it pincered open like an old-fashioned camera lens.

"You knew he'd have to improve on your garage doors, didn't you?" Toby said, impressed despite himself. Within minutes they processed through the second airlock and were landing beside Brent's red van on a beautifully flat and almost glowing marble landing pad, lit as it was by three powerful floodlights that had automatically turned on.

"Looks like he's carving it out one brick at a time." K'Narf said, nodding at the large pile of stone blocks stacked at the edge of the landing pad.

"Those are not just bricks, they are blocks of very pretty marble." T.J. said, looking carefully at the stacked blocks. "Those could be sold." She added, instantly knowing why Brent was cutting them out with such precision.

"Sold? Marble?" K'Narf repeated as he opened his door and got out. Moving quickly around to open the door for T.J. and helping her out, leaving Toby to get out as best he could.

"Whom do you sell them to?" T.J. asked, as Brent and DeeDee walked over to join them.

"Sharp as usual." Brent said, giving T.J. a slight bow. "Firm in Italy, they think they're coming from a quarry in China. Get a real nice price for them too." He added, answering T.J's question.

"How big do you plan to make this place?" K'Narf asked, looking around the large chamber Brent had hollowed out."

"Actually, it's already a little bit bigger than I had originally planned due to the selling price of marble." Brent said, pointing over to where he was still actively cutting out marble blocks. "Over there is my power room and that is my communications room, office and living quarters." He added, pointing at an entrance some fifty feet to the right of the landing pad.

"Communications room?" Toby asked, wondering why Brent needed a communication room.

"Yes, I have modified six large satellite disks and spaced them at equal distances around the outer surface of Marble Mountain, so I can have continuous contact with my antenna asteroid even though Marble Mountain is continuously rotating. And, if you haven't noticed, I set her up with the same revolutions per minute as the Whale, that way anyone coming over from the Whale would be comfortable here."

"Marble Mountain? Antenna asteroid?" Toby asked, the new terms throwing him.

"Marble Mountain is what I call this place, and my antenna asteroid is an asteroid about five miles from here that is extremely stable, I mean to the degree of zero movement, where I have set up receiving antennas. I have one aimed at Pluto, Earth and the Whale as well as all the open space where I seeded those solar-powered directional tracking relays I told you about." Brent said, beginning to realize Toby and K'Narf's ignorance at the extent of his operations.

"Do you mind showing us what you are talking about?" K'Narf asked, looking toward what Brent had called his communications room, office and living quarters."

"No, If DeeDee doesn't have any more questions. I've already told her about another asteroid approximately a hundred miles from here that is solid granite all the way through, not a crack or fissure in it. I had already x-rayed it and was going to use it myself until I found Marble Mountain. It should make an outstanding location for a warehouse. Exactly what she seems to need I think." Brent said, nodding at DeeDee.

"I may have more questions later, but for the moment I'm satisfied." DeeDee said, looking at K'Narf.

"Good, then let's go look at my digs." Brent said, as he turned and started walking toward the darkened entrance that had been cut into a solid wall of beautiful off-white marble that seemed to have a hint of purple shimmering in it. "By the way, most of the equipment your going to see can be transferred to the conference room in the Whale so you can track everything from there once the battle actually starts." Brent added, as he hit a light switch and lit up his digs as he called them.

"Don't you mean '*so we can track everything*'?" K'Narf asked, emphasizing his words as a way of reminding Brent about his sworn promise of not getting directly involved in the battle to come, as he followed him into a well lit disaster of a room that brought a gasp from both T.J. and DeeDee. One whole wall was lined with four drawer file cabinets labeled from A through Z with piles of unfiled paperwork on top of each of them. Over half a dozen large flat screens were hanging on the far wall with a maze of cables strung across the floor to a haphazard stack of computer components, all seeming to surround, more or less in the center of the room, an easy chair, a small refrigerator and a mattress spread over with a bright Indian style blanket and two large pillows that showed definite signs of having been slept on.

"Sorry about the mess, haven't been here in almost three months." Brent said, picking something up off of the floor next to his mattress and tossing it at a large grey garbage can.

"Brent!" T.J. said, scolding him with the tone of her voice as she stepped over to the nearest filing cabinet to see what it was that obviously needed filing.

"Careful, those are all in order, I just haven't had time to file them yet." Brent said, instantly trying to defend himself.

"What are these?" T.J. asked, holding up a studio quality eight by ten portrait of a young couple dressed in distinctive Snark blue-green spacesuits holding bubble glass helmets and smiling at the camera. It was obviously a picture of one of Brent's Teardrop fighter teams and it had a large sealed manila envelope fastened to the back of it with an oversized paperclip. Noticing as she asked the question that all the stacks were the same, pictures were of different fighter teams, but each of them had the same large manila envelope fastened to the back of it.

"Those are for my *'in case'* files." Brent said, lowering his voice to almost a whisper.

"In case files?" Toby asked, noticing that T.J. and DeeDee both seemed to know instantly what Brent was talking about.

"Yes, in case they don't make it. The pictures are for Debbie so she will be able to paint a portrait of the lost team for the chapel over in the Whale." Brent said, remembering the painting of Wonder that hung there now.

"And the envelope?" T.J. asked, carefully replacing the picture and its manila envelope back where she had picked it up.

"The envelope contains letters to their loved ones to be delivered only, and I stress only, if they are lost in the coming battle." Brent said, noticing the effect his words were having on DeeDee.

"That's kind of morbid, isn't it?" DeeDee said, grimacing.

"Yes, you might say that, but the U.S. does the same thing with their soldiers when they know they are going to be serving in a combat zone."

"And these are all bonded couples?" DeeDee asked, waving at all the pictures stacked on top of the file cabinets. "Mind if I ask why they all have to be bonded couples?" She asked, looking at Brent.

"Do you know what a bonded couple is?" Brent asked, looking back at her.

"Same thing as being married I guess."

"Not exactly. Correct me if I'm wrong." Brent said, looking at T.J. "But when God bonds two people he allows them to see everything that has ever happened to the other person, whether it's good or bad. Then he makes it so you can feel the feelings of each other. You know if your bonded mate is happy, sad, frightened or even angry. Is that about correct?" Brent asked, looking at T.J. for his answer.

"In a very crude way, that does explain it." T.J. said, remembering the emotions she had felt when God had bonded her to K'Narf.

"Then tell me Toby, what would happen to you if you lost Irena?" Brent asked, turning his attention to Toby.

"I think I'd die." Toby said, the question catching him by surprise. "I mean, I don't know how I'd go on living, I'd be devastated emotionally." He added, trying to think what it would be like to go on living without her.

"Exactly, and if we get into a major battle, and my fighter crews are made up of unbonded teams, I could wind up with devastated widows or widowers all over the place. If they're bonded they both go out together." Brent said, looking back at DeeDee. "And there is the simple fact that bonded couples make better teams."

"Better Teams?" DeeDee asked, beginning to realize that Brent had good reasons for insisting that all his fighter teams be bonded couples.

"Yes, while one of them is concentrating on flying, the other one is concentrating on firing their weapons. Being able to instantly feel each other's feelings gives them an edge that an unbonded team would not have. An edge that may very well save their lives in a combat situation." Brent said, looking at DeeDee to see if she was agreeing or not with his reasoning.

"And yet you are not bonded, why?" DeeDee asked, thinking it a reasonable question.

"I have made the decision to remain as I am until we have secured our Solar system." Brent said, noticing the amused look on T.J.'s face.

"I'm glad Redwing did not know that. I would never had talked her into moving up to the Whale to help out." T.J. said, noticing the surprised look on Brent's face at the mention of Redwing.

"What do you mean grandmother?" Brent asked, using the term of endearment for T.J. that dated back to the time before she had learned to change her appearance.

"Did you not notice her princess outfit, and the beaded headband that spoke of her linage back to Chief Joseph? The same linage you are so proud of?"

"Yes, grandmother, but I..."

"She took the position in the Whale." T.J. said, cutting Brent off. "Because she knew that it would give her an opportunity to meet you. She had heard very much about you, and was very excited when she heard that you were coming to the Whale for a meeting. Of course that was until she found out that you were the nasty little boy that had caused her to be sent to an all girls school for what she called '*eight years in a white girls purgatory*'."

"She wanted to meet me?" Brent asked, not quite believing what T.J. had told him.

"Yes, Little Eagle, she really did." T.J. said, using Brent's Nez Perce name.

"Mister Conrad." DeeDee said, breaking the moment between T.J. and Brent that went back to the time when she had been his mentor. "I was wrong about you're insisting that all of your fighter teams be bonded couples; I shouldn't have argued or gotten angry with you without knowing the facts.

Will you accept my apology?" She asked, stumbling over her words a little as she said them, obviously embarrassed.

"Did God forgive you for getting angry at Him?" Brent asked, relieved at the change of subject.

"Yes, of course He did." DeeDee said, surprised by Brent's question.

"Then I guess I have no choice but to forgive you as well." Brent said, openly grinning at DeeDee's embarrassment.

"You're laughing at me! I'm apologizing and you're laughing at me, why I ought to..."

"DeeDee!" T.J. exclaimed, stopping DeeDee from finishing what she was going say.

"Miss DeeDee." Toby said, grinning. "Are you, by any chance, feeling an overpowering desire to grab Mister Conrad by the nap of the neck and throttle him?" He asked, jumping into the fray as he looked at DeeDee and nodded at Brent.

"Well, I wouldn't go so far as to throttle him, but I wouldn't mind slapping that grin off his face." DeeDee said, seeing Brent's grin get even bigger. It was obvious that he was tickled about being able to yank her chain.

"In that case, welcome to the club. Everyone on staff has had that identical feeling at one time or the other; so if you find yourself in a situation where you need assistance in any form of retaliation or revenge, please feel free to call on any of us." Toby said, smiling at DeeDee and giving Brent a mock bow of honor.

"Brent, I don't know how you do it." K'Narf said, grinning in spite of himself. "But we need to get going, and you and Minh Le still need to figure out how to block that freezing attack you told us about.

"Okay, but let me check something first." Brent said, turning around and half jumping into his easy chair and pulling out two of the many controllers sticking out of a side pocket on the chair. Almost instantly one of the large flat screens against the wall lit up showing the downed mother ship out on Pluto. "Hey look, I got one!" Brent cried out, zooming in on a round shiny object that looked like a big ball bearing lying on the ground.

"Got one what?" K'Narf and Toby both asked.

"One of those Snark probes." Brent said, grinning up at them.

"Snark Probes?"

"Yes, do you remember when those lights appeared over the Temple Mount? It was about two in the morning, they were caught on video from at least four different angles, remember." Brent said, enlarging the screen a little more as he zoomed in closer to the downed probe.

"Yes, I remember reading about it, why?"

"Well, that's one of them. I watched those videos and it was obvious from the way they were moving that nothing living could survive moves like that without being smeared against a bulkhead, so they had to be robotic in nature. If they were robotic then it stood to reason that they were probes sent here by the Snarks. If that was the case, I also figured they would sooner or later check out their mine out on Pluto so I set a trap for them." Brent said, standing up and walking over to the flat screen to get a closer look.

"Trap, what kind of trap?"

"I set up four quarter inch lasers to automatically fire if one of them came down to look at the damage where you cut the head off of that monster." Brent said, turning and giving K'Narf a nod. "And it worked." He added, turning back to the flat screen.

"Automatic fire? What if one of our people had gone out there to get something and flew into your trap, what then?" K'Narf asked, seeing what he thought was a serious flaw in what Brent had done.

"Lasers wouldn't have fired." Brent said, still studying the flat screen, trying to get a measure on the size of the downed probe.

"Why not?" Toby asked, jumping into the conversation.

"Because everything we have that flys has been tagged." Brent said, using his fingers to measure with.

"Tagged?" K'Narf asked. "What do you mean tagged?"

"Go out and check behind the Lady's front bumper and you'll find a small flat signaling disk about the size of a silver dollar super glued there, and another one just like it glued to the inside of the back bumper. None of our weapons, laser or missiles, will ever fire at or hit her as long as she has those I.D. disks glued to her. There will be no deaths from friendly fire on my watch." Brent said, turning to look at K'Narf and then at Toby. "I think I can get this thing into one of our vans." He added, looking again at the flat screen.

"When did you..."

"Would you excuse me for a minute?" Brent asked, cutting K'Narf off as he closed his eyes for what seemed like minutes, though it was actually less than one. "Good." He said, nodding his head as he opened his eyes. "I need to borrow your Lady, if you don't mind."

"What for?" K'Narf asked, knowing Brent had been flashing someone.

"Dan and Susan Ripplinger are going to meet me out at Pluto and help me load that thing, but I'd rather use the Lady with her new disappearing paint job than my van. They'll be flying one of the new fighters that Glenn and his guys just finished painting the same way. That way we will both be invisible if another one of those probes show up." Brent said, finishing his explanation.

"That's a long trip, do you have any food here?" T.J. asked, as she squatted down and opened Brent's little icebox. "Seven bottles of Coke and half a box of Snicker candy bars."

"And one orange." Toby said, looking over T.J.'s shoulder.

"Scratch the orange." T.J. said, as she picked it up and tossed it over her shoulder to Toby.

"This thing is harder than a rock." Toby said, catching it and deftly tossing it at the garbage can.

"K'Narf, what do we have in the ice chest?" T.J. asked, referring to the Coleman cooler that they kept in the Lady.

"Some bottled water, three bananas and half a package of double stuffed Oreo's." K'Narf said, mentally double-checking the inventory of their little cooler.

"Not much for what, three days out and four back to Earth?" T.J. said, guessing at the distance.

"About that." Brent said, getting a box from the corner to put the Cokes and candy bars in.

"Maybe Dan and Susan will share some of theirs." He added, handing the box to T.J.

"Not if they're smart." Toby said, grinning at Brent.

"You must learn to sleep with one eye open." Brent fired back, giving Toby a downright nasty grin.

"What's the code to get through your airlocks?" K'Narf asked, thinking he'd hate to get stuck in here and not know the code to get out.

"Same as the Whales." Brent said, noticing the surprised look on K'Narf's face. "Figured this place wouldn't be much help to you if I had you locked out in an emergency." He added, flashing K'Narf the same cartoon caricature that K'Narf had flashed to him at Toby's party when they had first met, the one that had K'Narf standing on the whale yelling "help" at a similar caricature of himself standing on another asteroid.

"Good, you wouldn't mind then if Redwing and I came over here and cleaned up all this filing for you, would you?" T.J. asked, changing the subject as she let Brent know that that was exactly what she intended to do.

"Redwing, would that be wise grandmother?" Brent asked, looking a little nervous.

"I will not let her put itching powder under your blankets." T.J. said, laughing at Brent's uneasiness.

"What about extra seats, we'll need two." K'Narf said, wondering if Brent had extra seats for his van, being a GMC, the seats out of the Lady wouldn't work.

"She's already rigged out with four, I've been hauling around some of my teams." Brent said, seeing the relief on K'Narf's face.

"Then let's get out of here, we have a lot to do." K'Narf said, stepping over Brent's mattress and helping T.J. up.

Chapter Five

Watching the airlock door pincer shut behind him in his rear view mirrors, K'Narf looked forward just in time to see the Lady shimmer and disappear, heading for her rendezvous on Pluto. "He's off." K'Narf said, thinking of his double stuff Oreos going with him.

"He's a very unusual young man, isn't he?" DeeDee asked, sitting in the back seat behind K'Narf. "And I think you are very fond of him." She added, looking at T.J.

"We all are." K'Narf said, leaning slightly so he could see DeeDee in his rearview mirror. "We may not always show it, but we all are and we're very grateful for everything he's doing. Our fate would be much less certain without him." He added, trying to make sure the DeeDee truly knew that they all held Brent in very high esteem.

"How did you find him?" DeeDee asked. "And why does he call you grandmother?" She added, looking at T.J. as K'Narf started increasing their speed as they left the asteroid belt and headed for Earth.

"Because for many years, I was." T.J. said, smiling at the memory. "The Lord touched him and removed his filters right after he turned fourteen." She added, remembering how she had first met Brent.

"Filters?"

"It's just an analogy, but it's the best way we've found to explain it. Each one of us has been touched by the Lord in a way that removed the barriers, or filters as I like to call them, between our conscious and subconscious minds. Science has long known that as normal humans we use only a very small percentage of the brains God gave us, and now after our Lord's touch, we seem to be able to use them all." T.J. said, explaining her little analogy for the hundredth time.

"I'm sorry, go on." DeeDee said, realizing she was interrupting the flow of T.J.'s story.

"Yes do, I've never heard how you found him either." K'Narf said, adding his curiosity to DeeDee's.

"Me too." Toby said, chiming in.

"Okay, okay." T.J. said, warming up to her captive audience. They had seven hours to go before getting home.

"Like I said, he was just fourteen when the Lord touched him, and he was just as shrewd, smart and innovative then as he is now. He knew that he couldn't get anywhere near the gaming tables in Vegas as a fourteen year old, so he made himself an elaborate outfit and pretended to be Saud Royalty. He also dressed up two of his larger cousins as bodyguards. He had the entire staff of the Grand bowing and scraping to him when I first saw him." T.J. said, trying to smile at the memory as the gravity increased to the three-point-seven G's K'Narf had set for their return journey.

"Saud Royalty, you mean like Saudi Arabia?" Toby asked, surprised.

"Yes, and he pulled it off. He was smart enough to know not to bring attention to himself as a big winner, so he started using shills just like K'Narf did when he came to Vegas and he made the same mistake K'Narf did, he tried to use me." T.J. said, again trying to smile against the pressure of constant acceleration.

"What happened then?" DeeDee asked, wanting T.J. to continue her story.

"I confronted him. At first he tried to maintain his absurd sham, then broke down and told me everything. I took him under my wing so to speak, became his new '*grandmother*' as he like to call me, and with my help he quickly learned how to play the stock market and the commodities market. By the time he was sixteen he was majority stockholder in three rather decent sized companies, not counting a very respectable mixed stock portfolio." T.J. said, obviously very proud of her protégée.

"How old is he now?" DeeDee asked, beginning to understand the relationship between T.J. and Brent.

"He just recently turned nineteen." T.J. said, trying to adjust to the new seat, it didn't feel as comfortable as her old seat in the Lady. "Technically he's still a teenager."

"You mean obnoxious teenager, don't you?" Toby asked, more seriously than he intended.

"Are you saying that all teenagers are obnoxious, or just Brent?" T.J. asked, turning in her seat enough to wink at DeeDee.

"Why?" Toby asked, sensing he was being set-up.

"Well, since K'Narf is also nineteen, then he too is technically still a teenager." T.J. said, trying to bait Toby into calling K'Narf obnoxious.

"K'Narf was born old, I don't think he was ever a teenager." Toby said, thinking fast as he tried to avoid T.J.'s trap.

"So now you're calling my husband an old man, is that right?"

"No, I'm calling Brent an obnoxious teenager, that's all I'm doing and you know I'm right T.J., so stop it." Toby said, knowing when he had been bested.

"Okay, I'll admit that he is alternating between, as you say, an obnoxious teenager and an adult. We must give him a little more time. I think Redwing might help with his growing up." T.J. said, smiling to herself as she let Toby off the hook.

"That's easy for you to say, you don't have to sleep with one eye open." Toby said, complaining. "You don't think he's actually going to hurt me do you?" He added, wishing he could see the expression on T.J.'s face.

"I think you should not try to one-up him all the time. You will always lose you know."

"I know, but when he smarts off, I just can't seem to help myself." Toby said, realizing what T.J. had said was the truth; he really wasn't up to matching wits with Brent.

"Remember, you are the adult, and now you must learn to sleep with one eye open." T.J. said, grinning even though her face didn't show it.

"Speaking of sleep, I advise us all to get some. We're all going to have to hit the ground running when we get home, and Toby this may be the last time in a long time that you will be able to relax and sleep with both eyes closed." K'Narf said, unable to resist picking on Toby a little himself.

"When we get home it will be Friday, and the sun will be setting within hours of our arrival." T.J. said, reminding K'Narf about their need to honor the Sabbath.

"How much time will we have?"

"A little over three hours, why?" T.J. asked, carefully turning her head to look at K'Narf as the continued pressure of the van's thrust seemed to push her deeper into her seat.

"Good, that will give me time to check on the factory and for us to stop by the school and get ourselves updated." K'Narf said, mapping out the time they would have so none of it would be wasted.

"That is what you must do, I must go to the store or you will find yourself fasting during Sabbath." T.J. said, wiggling even deeper into her seat.

"Okay." K'Narf said, knowing from experience not to argue. "If you'll set your alarm for six-and-a-half hours, let's try to get some sleep." He added, nodding at T.J. as the van settled down to a steady pulsing thrust.

"Don't you know how to set a mental alarm?" DeeDee asked, surprised that K'narf would ask T.J. to do something so simple.

"Yes, but I consider that woman's work." K'Narf said, flashing a wink at T.J., knowing full well he was yanking DeeDee's feminist chain.

"A woman's what!"

"It's all right DeeDee." T.J. said. "Keep in mind that I'm the one that gets to wake him up!" She added, as she flashed DeeDee a picture of her pouring a bucket of ice water over K'Narf's sleeping form complete with ice cubes.

"Woman's work." DeeDee repeated, calming down as she realized that K'Narf was teasing her. "Good night." She added, as she closed her eyes.

"Good night DeeDee." K'Narf said, closing his own eyes.

"Did you get any sleep?" K'Narf asked, flashing his question at T.J. as the van finally slowed down enough for him to turn it around so he could see the approaching Earth through the windshield instead of through the big side mirrors. Flying backwards during deceleration had always bothered him.

"No, I found it just as impossible as you did." T.J. flashed back, stretching for the first time in a little over seven hours.

"My grandfather snored, but compared to DeeDee he was a rank amateur."

"When she repairs her face, she should be able to fix her adenoids also." T.J. flashed in a vain attempt to defend DeeDee who was still doing a credible imitation of a chain saw.

"How can Toby sleep through that, anyway?" K'Narf asked, flashing the question as he looked at the peacefully sleeping Toby in his rearview mirror.

"He told me once that during the war his company was on the receiving end of an artillery barrage. He knew there would be no ground action until it lifted, so he curled up in his foxhole and went to sleep."

"During an incoming artillery barrage?" K'Narf asked, looking again at the sleeping Toby.

"Soldiers learn to sleep when and where ever the opportunity presents itself." T.J. flashed. "This van does not have the ability to be invisible, does it not?" Realizing, as K'Narf suddenly did, that the area where the factory was located was still lit up by the sun.

"Guess I'll have to get updated later." K'Narf flashed, knowing that they wouldn't be able to land until it got dark enough to drop down undetected.

"I will flash Sylvia and ask her to shop for some fresh fruit for us. She can leave it on our doorstep." T.J. flashed, mentally dialing up Sylvia's frequency. "She will be happy to." T.J. flashed moments later.

"Good, hope she gets us some of those blood oranges."

"It will be Sabbath when we land, I think we ought to ask DeeDee to spend it with us." T.J. flashed, knowing what K'Narf's reaction would be before she asked the question.

"Sabbath is for resting, how do you plan to sleep?"

"I have a very nice pair of ear muffs." T.J. flashed, grinning.

"What about me?"

"You must learn to sleep like Toby." T.J. flashed, grinning at the look on K'Narf's face. "No, seriously, we will put her up in the downstairs guest room. The one just off the kitchen, we should not hear her from there."

"If you say so, shall we wake them up?" K'Narf asked, this time speaking his words instead of flashing them.

"Yes, very much yes." T.J. said, speaking also as K'Narf reached back and shook Toby's knee just as DeeDee gave out a loud snort.

Chapter Six

"They're ready for you sir." The young man said, sticking his head into K'Narf's office as he delivered the message.

"Thank you, tell them I'll be right there." K'narf answered, clicking on the apple and putting his iMac to sleep. Then, getting up and heading down the stairs toward the manufacturing floor, he noticed how it had changed in the last few weeks. No longer did it have the look of randomized chaos, it now looked like a very well organized assembly line, which is exactly what it was.

"If you'll step over here to the changing room, sir, we'll get you suited up so to speak." Another young man said, as he greeted K'Narf at the bottom of the stairs.

"What do you mean 'so to speak'?'" K'Narf asked, noticing the flicker of a grin on the young man's face. "And what is your name?" He added, following him into a small hastily put together changing room.

"David, sir, David Ben Ryan." David said, as he pointed to the outfit he wanted K'Narf to put on.

"You've got to be kidding." K'Narf said, looking at what appeared to be brightly colored blue, red and black set of silk tights.

"We don't have a Snark flight suit that will fit you sir, so we bought you these." David explained, openly grinning now. "They're bike racing tights." He added, explaining what the original purpose of the spandex garment was.

"And exactly why do I have to wear this?" K'Narf asked, taking it off the hook it was hanging on and pulling at the stretchy fabric.

"The cockpit has been designed to apply pressure to the body so it cannot be tossed around during, shall we say, violent maneuvers of the fighter. Wearing a belt, watch, wallet, keys or even the thicker seams in your clothing could cause serious bruising, sir."

"You sure none of your Snark suits will fit me?"

"No sir, we could grow you one, but it would take about a week." David said, measuring K'Narf's size and bulk with a practiced eye.

"Grow me one, what do you mean, grow me one?"

"I thought you knew, sir. Mister Conrad's people figured out that the reason those Snark suits worked so well was because they were made up of a living organism. That's how they're able to convert carbon dioxide back into oxygen as fast as it's breathed out, and how they can absorb all bodily waste and recycle it the way they do." David explained, surprised that K'Narf didn't know about the Snark space suits.

"You really grow them, how do you do that? K'narf asked, surprised by David's answer.

"Yes sir, you create a mannequin that's the exact size and shape of the pilots body, and then grow the suit on it. Of course it's fitted out with a helmet ring and hermetic zipper seals so the pilot can get in and out of it, and of course you do know that it only functions if it is actually touching the pilots skin." David said, embarrassing himself a little.

"You mean the pilots have to be naked when they put it on?" K'Narf asked, thinking for the hundredth time he'd like to strangle Brent, but then to be fair I haven't been over to Premo's in over a week to get updated, and the growing of the suits might be in the latest update, he thought. "What about a helmet?" He added as he put the growing of the suits out of his mind.

"You won't need one for just a short test ride, sir.

"And I really have to put this on?" K'Narf asked, hoping against hope that David would say no.

"Yes sir, that biking outfit is the closest thing we could come up with that wouldn't cause you to get bruised by the pressure system in the cockpit." David added, as he stepped out of the small changing room and let the curtain drop giving K'Narf full privacy.

"You, uh, look very, uh, nice, sir." David said, stammering as he tried not to laugh. K'Narf's size and muscles had stretched the fabric as far as it was possible to stretch it. "Those colors are really your colors too, sir." He added, openly grinning now.

"I hear DeeDee is looking for a swamper to help out on her next run." K'Narf said, surprised by the immediate effect his words had on David.

"I'm sorry sir, I, I really am. I didn't mean any disrespect." David stammered, coming to what amounted to full attention if he had been in the military.

"Take it easy, I was just kidding." K'Narf said, wondering why DeeDee's name had had such an affect. "Is she really that bad?"

"Uh, well, see that guy over there with the red baseball cap on?" David asked, pointing at the man he had just described. "His name is Joey Johnson and she tagged him to go with her on her last run."

"Tagged him?"

"Uh, yes sir. She tags someone, and they're it, kind of like conscripted labor, sir. Anyway, she tagged Joey and when they were unloading up at that warehouse of hers he got the tomatoes and tomato juice mixed up, or something like that. She got upset and flashed a picture of Joey on a gurney in some kind of hospital room. He was all busted up and bloody and screaming in pain. I don't think he's got that image out of his mind yet, sir." David finished, still standing at attention and not smiling.

"I see, you know she spent forty-two years working in a hospital emergency room. That's probably where she got the image she plugged Joey into."

"That kind'a makes sense, sir, I guess."

"Would it help if I spoke to her about it?" K'Narf asked, seeing David relax a little.

"Yes, sir, anything you could do would help. I mean we don't hate her or anything like that, most of us think she looks like Maureen O'Hara."

"Maureen O'Hara?" K'Narf asked, remembering that DeeDee had said she used to look like a person called 'Maureen O'Hara'"

"Movie star, sir. We're all just waiting for the Duke to come along and put her over his knee and spank her, uh, sir." David said, immediately thinking he'd said too much. "Uh, they're waiting for you sir." He added, pointing at three men standing by one of the new fighters.

"Okay." K'Narf said, as he and David walked toward the waiting men leaving K'Narf to wonder who the Duke was. Have to ask Toby when I get a chance, he thought as they approached the men and the sparkling, jewel like fighter. Noticing at the same time the stares he was getting from all over the assembly floor.

"Mister K'Narf, this is Karl Grauer, our head designer, Mitch Murphy his assistant and Peter Vanlieu your pilot." David said, introducing each man in turn.

"Karl, it's nice to see you again." K'Narf said, remembering the time he had spent working with Karl on the original three fighter's they had built as he reached out and shook Karl's hand. "And Peter, you were one of the pilots with Brent at that debriefing meeting." He added, shaking Peter's hand as well.

"Yes sir." Peter said, as K'Narf released his hand.

"And this is my right hand and sometimes my left." Karl said, pointing at Mitch Murphy. "I don't know what I'd do without him." He added, as K'Narf shook hands with Mitch.

"It's my pleasure gentlemen. Well Karl, you said if I gave you a little time you could come up with a design that would knock my socks off and it looks like you did exactly that." K'Narf said, looking at the sparkling Teardrop fighter sitting quietly on it's cradle.

"She's a sweet one all right. Twenty-two-and-a-half-feet long and nine thousand, five hundred pounds of pure nasty." Karl said, turning to look at his creation.

"How nasty?" K'Narf asked, noticing she seemed to be faceted like a finely cut diamond.

"All those facets make her invisible to radar. Each facet is a half inch of intensified glass, bullet proof glass to you, and the mirror backings gives her the ability to shatter and deflect a laser beam hit as long as it doesn't hit square on. On top of that she has two twin half-inch laser cannons and carries a full complement of thirty missiles, six aimed at anything on her tail and twenty-four aimed at anything she's chasing. All of that, coupled with this new paint and camera combination that makes her literally invisible to the enemy." Karl said, extremely, and obviously very proud, of his Teardrop fighter.

"What do you call her?" K'Narf asked, thinking every ship should have a name.

"She's TD3, meaning she's the third one off the line. We decided that since we were going to be making so many of them that it would get too confusing to try and give them individual names, so we're just calling them by the number of their birth, so to speak." Karl said, turning back to look at K'narf as he was speaking.

"And her drive cores?" K'Narf asked, trying to picture how the cores had been aligned inside her.

"Right here, see how she tapers down to about six feet across right here." Karl said, indicating a spot about nine feet from her sharply pointed tail. "There's a heavy duty six foot drive core embedded in there that has more than double the kick of those clusters of five two foot cores they use as the back drive for the vans."

"Wait a minute, you're saying that your six foot drive core is equivalent to ten two foot cores?" K'Narf asked, quickly doing the math.

"More than that when her afterburner kicks in." Karl said, nodding his head emphatically.

"Afterburner, how do you put an afterburner on a core?"

"See this long tapered tail?" Karl asked, pointing at the final nine feet of the Teardrop's fuselage. "Well, that happens to house a very powerful tapered electromagnetic rod. When energized it pulls the cores thrust cone down into a concentrated beam instead of a dispersing fan of energy which almost doubles the cores drive force." Karl explained, even prouder if that was possible, of his afterburner as he called it.

"This I have to see to believe." K'Narf said, sounding a bit dubious.

"Actually, if you will allow us to fit you into the aft cockpit, I will arrange for you to feel it which is a tad bit better than seeing it." Peter said, pointing at the ladder that led up to the aft open cockpit hatch. "You did say you wanted to take a test flight as soon as it was possible to arrange it, didn't you?" He asked, seeing a fleeting look of hesitation on K'Narf's face.

"How about side to side movement?" K'Narf asked, as he put his foot on the first rung of the ladder.

"She has a ring of nine two foot cores circling her midsection allowing her to jump in any direction necessary to avoid getting hit, or to pursue an enemy fighter." Karl said, jumping back into the conversation as K'Narf started to climb the ladder.

"Still think you can produce one every three days?" K'Narf asked, as he climbed up the ladder with Peter close behind him.

"Yes, if Toby, I mean Mister Zebulon, can keep the materials coming. We have set-up rotating shifts of six hours each going twenty-four hours a day for all six days of each week. Unless you can get us another factory or expand this one, that's about the best we can do." Karl said, as K'Narf reached the top of the ladder and looked into the cockpit.

"This thing looks like an open bag of marshmallows." K'Narf said, looking at the interior of the cockpit.

"Very apt description." Peter agreed. "Just wait until we inflate them." He added, grinning as he helped K'Narf into the cockpit and settled him into the seat.

"Inflate them?" K'Narf asked, as he allowed Peter to put his hands into the two holes in front of him.

"Yes, now sit very still as I do just that." Peter said, as he pushed a big green button on the console in front of K'Narf. Instantly the marshmallows as K'Narf had called them began to inflate.

"I hope I don't get an itch on my nose." K'Narf said, realizing that he was being imprisoned by hundreds of miniature pressurized airbags. "If you need more pressure just hold down the green button, if you want to release pressure press the red one." Peter explained, still holding his finger on the green button.

"It's getting awful tight." K'Narf said, realizing the amount of pressure on his body was getting uncomfortable.

"Okay." Peter said, taking his finger off the green button, "But if you feel yourself being thrown around, just hit the green button for a few more seconds, okay." He added, as he stepped back and closed the hatch over K'Narf's head. Within minutes K'Narf felt as well as saw the TD3 lift off its cradle and start a slow turn as it headed for the open door in bay one.

"I'm going to cloak her now." Peter flashed, using the common frequency so everyone who wanted to could listen in on their conversation.

"You call it cloaking, too?" K'Narf asked, recognizing the same wording he had used.

"Yes, Brent, I mean Mister Conrad, said that's what you called it, cloaking and de-cloaking." Peter said, surprised by K'Narf's question.

"He did, did he? Well, actually it was Toby who said it. Something about Klingons doing it, or something like that." K'Narf flashed back.

"Yes, they cloaked and de-cloaked their Warbirds didn't they." Peter flashed, showing everyone a picture of a strange looking ship that K'Narf assumed was a Warbird as it shimmered and faded out of sight. "Ready?" He added, flashing K'Narf the question as the TD3 cleared the bay door and started lifting straight up.

"Where are we going?" K'Narf asked as he was pressed first down into his seat and then into the back of it.

"I'm going to stay low until we get over Alaska. I don't like lifting through the satellite belt, too dangerous." Peter said, suddenly throwing K'Narf violently to his left as he made a sharp turn to his right.

"I think I need more pressure." K'Narf said, moving his right hand just enough so that he could touch the green button; pressing it down until he felt there was enough pressure to hold him steady if Peter made another sharp turn like that last one. "What are the screens and the joystick for?" He asked, looking at the console in front of him.

"Are you right or left handed?"

"Right, why?"

"Good, that cockpit is made for my wife, Antoinette, and she's right handed too." Peter said, flashing K'Narf a picture of his wife. "The Joystick is for tracking and firing the laser cannons." He added, as the green forest of Canada began to appear on the right hand screen in front of K'Narf.

"Tracking, they're moveable?" K'Narf asked, surprised.

"Yes, about thirty degrees. If I can get a target onto the right hand screen in front of you, you should be able to zero in and hit it with a laser cannon using your joystick." Peter explained, climbing a little as snow covered mountain peaks appeared in front of them.

"And the left hand screen?"

"Shows what's coming up behind you. The purple button on the left side fires one of the six rear missiles each time you push it." Peter explained. "And the purple button on top of your joystick fires one of the twenty-four missiles from the forward launchers each time you push it."

"And the two big zero's on the upper left of each screen?" K'Narf asked, beginning to figure out how the TD3's defense systems worked.

"It means you're out of missiles, we're not armed at the moment."

"You don't trust me?" K'Narf asked, only half kidding.

"Teardrop missiles are packed with the most powerful explosive known to man. It would be a serious disaster if one of them were to be accidentally launched regardless of where it hit." Peter flashed, not denying K'Narf's statement about not trusting him.

"You mentioned the left and right hands. What was that all about?" K'Narf flashed, changing the subject.

"A right handed person is inherently faster and more adroit with their right hand than their left so we put the joystick controlling the laser cannons and the forward missile launchers on their right. Their left hand is given the task of pushing buttons and firing the rear missile launchers only. Of course we reverse the console for a left handed person." Peter explained. "Okay, are you ready to play co-pilot and see what our little Teardrop can do?" He asked, as K'Narf was suddenly thrown violently back and down in his seat as Peter headed for open space.

"That was quick." K'Narf flashed, realizing that they were already deep in the blackness of space.

"Yes, but you must realize that I cannot put her through her real paces because neither of us have compressed our bodies to take the tremendous G forces she would demand of us." Peter flashed, reminding K'Narf that this was a demonstration flight only and that it had been arranged especially for him because he had asked for it.

"Okay, show me what she can do." K'Narf flashed, bracing himself.

"Right, pretend there is a missile coming straight at us." Peter flashed, as he twisted his joystick hard to the right.

"I need more pressure!" K'Narf flashed, slapping at the green button, having been viciously slammed to the right into pressure pillows that felt so soft that they almost allowed him to hit the bulkhead. Moments later, with bruising pressure holding him from all sides, he was being brutally and violently tossed about as the TD3 spun, twisted, leaped and flew as though she was in a battle for her life. And this isn't the best she can do, K'Narf thought to himself.

Chapter Seven

"Are you all right sir?" David asked, having climbed up the ladder and opened K'Narf's cockpit hatch. Getting no answer he reached in and hit the red button and deflated the multiple airbags that had K'Narf immobilized. "Are you all right sir?" He asked again, repeating his question.

"I don't think there's a proper answer to that question." K'Narf said, letting out a low heartfelt groan as he tried to move various parts of his body.

"Can I help you out sir?" David asked, reaching in and taking K'Narf by the arm.

"That'll be the only way I'll get out." K'Narf said, wincing in pain as he tried to use his legs.

"Easy does it sir, watch your step now. These ladders can be a little slippery if you're not careful." David said, as he guided K'Narf's left foot to the bottom rung and then his right foot onto the floor, steadying him as K'Narf got his balance, grabbing and holding onto the ladder for support.

"That's the most awesome and painful experience I think I've ever had." K'Narf said, turning to face Karl. "Now, I've got a favor to ask of you." He added, as he looked first at Karl, then at Peter, Mitch and then David in turn.

"What is it?" Karl asked, speaking for all of them.

"There is no way for anyone to understand what your Teardrop fighter is capable of by reading reports. The only way a person could truly know her abilities is to take a test ride like the one you just gave me." K'Narf said, looking directly at Peter. "And I have a Chief of Staff that needs to understand those abilities as much, if not more so, than I do."

"What are you saying sir?" Peter asked, realizing that K'Narf had something specific in mind.

"I am going to get dressed and somehow get myself back up those stairs and behind my desk, then I'm going to call Mister Toby Zebulon into my office and lie through my teeth by telling him that I've just had the most thrilling and exhilarating experience of my life. I'm also going to tell him that I've asked you guys to hang around long enough to give him a quick test ride as well and that you are waiting for him."

"Not a problem sir." Peter said, grinning.

"That's the problem." K'Narf said, nodding at Peter. "If he sees any of you grinning like that, you won't be able to get him within twenty feet of that cockpit. I need you all to keep a straight face and act like this is nothing more than a little test run so he can get the feel of your Teardrop fighter."

"He doesn't have any physical problems, does he?" Karl asked, thinking there might be something wrong with Toby.

"No, Toby's an ex-Israeli Special Forces Commando, and he is suspicious of everything and everyone, that's all, but I want him to know the real capabilities of what you've created here. I received all the reports and nothing I read prepared me for what I went through in that cockpit. He needs to know what I know. It's as simple as that." K'Narf said, grinning in spite of himself. "Now, Mister David Ben Ryan, if you'll help me over to that changing room of yours I'll see if I can get out of this thing." He added, pulling at the spandex fabric covering his chest.

"I can rinse it out and run it through a quick dry cycle." David said, letting K'Narf know that he was fully on board with his plan to dry gulch Toby.

"Oh, K'Narf." Karl called out. "I got a message that your father called, wants you to call him first chance you get."

"Thanks, I'll do that." K'Narf said, as he stepped through the changing room curtain. Twenty minutes later he was sitting behind his desk talking to Toby.

"Yes, A young man named David is waiting for you at the bottom of the stairs. It will only take you a few minutes and I guarantee you won't be disappointed." K'Narf said, working hard to convince Toby to take a joy ride in one of the new Teardrop fighters. "They're taking valuable time out of their day just to accommodate us, so be sure you thank them for the ride. Oh, and be sure and pop your head back in here and tell me what you think of our new fighter before you go back to your office." He added, as he picked up his phone in an act of dismissal. Watching out of the corner of his eye as Toby left his office and turned right in the direction of the stairwell.

"Good!" K'Narf said, grimacing as he grabbed the edge of his desk and hoisted himself to his feet. "I need a cup of coffee and a sweet roll." He added, speaking out loud to no one in particular as he took a very careful step toward the door.

Ten minutes later he very carefully placed his coffee and a small Styrofoam plate holding a bear claw onto his desk, then cautiously sat back down, thinking he should call his father, then thinking that it could wait until he finished eating his bear claw. Several minutes later he picked up the phone and punched up his father's name and hit dial.

"K'Narf, thank you for returning my call so promptly." Joe senior said, having read his son's name on his phone's digital screen.

"My pleasure." K'Narf answered, wincing as he leaned back in his chair. "There's nothing wrong is there?" K'Narf asked, knowing that all of the required paperwork had been signed weeks ago along with a draft for the final payment.

"No, no, but when we last talked, you mentioned that if I should find another such property to let you know and...."

"You've found another factory town for us?" K'Narf asked, cutting his father off and trying not to let his excitement show in the tone of his voice. Trying in vain to emulate his father's ability of not showing emotion, especially in anything related to business.

"Yes, yes I have, but there is another interested party. Is it possible for you and your associate Mister Zebulon to drop by my office tomorrow say around ten? I could have one of my people on site with a video camera and we could watch a video stream of the site here in my office." Joe senior said, getting right to the point.

"Where?" K'Narf asked, excitement tingeing his voice despite his best efforts. Knowing as he did that they desperately needed a second location.

"Missouri, just outside of Kansas City. Not too terribly far from your current location." Joe senior said, knowing inherently that he had a sale.

"Yes, I know of no reason at the moment that we cannot be at your office by ten tomorrow; however, I will need to confirm the appointment with Mister Zebulon. If for any reason we cannot make it I will give you a call." K'Narf said, wondering privately if either him or Toby would be physically capable of making such an appointment.

"Thank you, I'll be looking forward to seeing you tomorrow then." Joe senior said, hanging up before K'Narf had a chance to ask him about his mother, or gramps and Junior.

"Well, that was short and...."

"You miserable, low down, sneaky!" Toby said, hanging onto K'Narf's doorjamb for support as he stuck his head in the door. "You need to pray that someone doesn't invent a time machine, because if they do I am going to go back in time and find your Great! Great! Great! Great! Great! Grandfather and make sure he doesn't survive the birthing process!" Toby exclaimed, groaning as he let go of the doorjamb and staggered across the room; crying out in pain as he dropped into one of K'Narf's desk chairs.

"Is there a problem?" K'Narf asked, trying to keep a straight face.

"Problem! I didn't think anything could be worse than what the IDF did to me during training, but I was wrong." Toby said, looking at K'Narf with wide-eyed disbelief. "And I thought you were my friend, someone I could trust!" He added, groaning again as he tried to shift into a more comfortable position.

"You needed to know..."

"You could have just told me!"

"No, there is nothing I or anyone else could have said, including supplying you with volumes of written reports, that could have told you what you learned about the capabilities of that Teardrop fighter during that one short test flight." K'Narf said, knowing he was right; knowing also that Toby knew he was right.

"You could have at least warned me."

"You wouldn't have gotten within twenty feet of that cockpit if I had." K'Narf said, grinning at Toby. "Anyway, we both have a rather serious problem that we need to solve by ten tomorrow morning." He added, seeing the immediate look of concern on Toby's face.

"What's that?"

"We have to be dressed in our best and at my father's office for a video streaming presentation about another possible factory site by that time tomorrow morning, and I don't think showing up in wheelchairs will help further our cause, not to mention what it would do to our image." K'Narf said, only half-teasing as he let Toby know of the appointment he had made for them.

"What were those pills T.J. gave us anyway?" Toby asked the next morning.

"I don't know, something she got from DeeDee." K'Narf said, feeling only slight twinges of pain as they got out of the taxi in front the building that housed his father's business and looked up at it. "I still don't like buildings with glass walls, they make me nervous." He added, stepping over to the big double glass doors and opening one of them for Toby.

"Thirty-second floor if I remember right." Toby said, as he punched the appropriate button. "How long do you think they'll last?" He added, feeling a twinge of pain in his shoulders as the elevator doors closed, giving him a feeling of being trapped.

"Not sure, we took them pretty early this morning though." K'Narf said, as the elevator started rising.

"Do you have any more of them?" Toby asked, as the elevator gathered speed in its upward rush.

"Yes, T.J. gave me six of them, why?" K'Narf asked, as the elevator abruptly slowed down, giving him a brief floating sensation.

"I want my share, that's all." Toby said, just as the elevator stopped and the door opened, thinking he'd like to take another one right now.

"Willetta sees us." Toby flashed, noticing that the formidable Willetta was getting up from her desk and actually coming around to meet them.

"Mister Etaguf and Mister Zebulon, Mister Fugate is expecting you." Willetta said, stepping over and opening the door that they both knew led to his father's office. "Please." She added, holding the door for them and smiling.

"Are we in the right building?" K'Narf asked, flashing Toby as he acknowledged Willetta with a nod and stepped through the door.

"I think this is what you might call the red carpet treatment." Toby flashed back, as a very pretty and very young secretary jumped up and came toward them. "Right this way gentlemen." She said, as she ushered them to her bosses' office. "Please." She added, as she opened the door for them.

"Ah, there you are." Joe senior said, as he got up from his desk and came around with his hand extended.

"Father, how are you?" K'Narf asked, shaking his hand; noticing the exceptional difference between this reception and the first time he and Toby had shown up here.

"Fine, fine. Please, take a seat over here. I've got a screen set-up and two of my associates are on site with cameras. You can take a preliminary tour of the factory and housing complex from here via video streaming." Joe senior said, shaking Toby's hand as well. Obviously pleased with himself and the demonstration he had set up. "Would you like a little refreshment?" He added, asking the question as he motioned to a small counter containing everything from hot coffee, Coke, orange juice and water along with a small assortment of sweet rolls.

"Like I said, the red carpet treatment!" Toby flashed, as he picked up a bottle of water and sat down.

"Thank you father." K'Narf said, also picking up a bottle of water and joining Toby at the small conference table.

"That pill is definitely wearing off!" Toby flashed. "That hurt just to set down." He added, flashing K'Narf again.

"I know, I felt definite shooting pains in my hips when I set down too!" K'Narf flashed back.

“Slip me another one of those pills when you get a chance.” Toby said, getting comfortable and opening his bottle of water.

“Okay, but don’t let my father see you take it.” K’Narf said, flashing Toby the warning as he too opened his bottle of water.

“What I have located for you is an unoccupied factory that was built less than ten years ago. It is in pristine shape with exactly two hundred and twenty-five thousand square feet of factory floor space as you can see here.” Joe senior said, bringing a picture of the factory on line. “My associate, as you can see, is standing in the middle of the factory floor where she is giving you a full circle view of the area itself.”

“Very impressive.” Toby said, remembering what the factory floor had looked like at the Sutton Wells site when they had bought it.

“She will now walk you through the factory office complex.” Joe senior said, settling back in his chair and watching the screen, giving K’Narf a chance to slip Toby a pill and take one himself. “There are four loading bays in the dock area; however this site does not have a rail line going to it.” He added, as the camera continued to walk around the office complex and show them the various offices and a very nice auditorium where large meetings could be held.

“I like the auditorium.” Toby said, noticing that K’Narf was being unusually quiet.

“Yes, it was very nice, wasn’t it, now let me take you to the housing complex.”

“Housing complex, don’t you mean town?” Toby asked, glancing sidewise at K’Narf.

"No, this site is not directly associated with a town, the owners instead built a small housing development they called a sub-division; it comes with the factory and is included in the purchase price. The nearest actual town is Kansas City which is some twenty-three miles distant." Joe senior said, turning to look at Toby and then at K'Narf. "Exactly what I was led to believe you wanted." He added, turning back to his screen after giving K'Narf a quizzical look.

"Yes, this is exactly what we are looking for." Toby said, realizing as he said it that this site met all of their criteria as the screen lit up with a picture of a classical tree lined street with very pretty houses lining both sides; houses that were obviously vacant.

"Good." Joe senior said, watching as the camera reached the end of the street and turned around for a final shot looking back up from the other end. "There." He added, as he put the screen to sleep and turned around to face K'Narf and Toby, looking at them expectantly.

"As Toby said, it's exactly what we're looking for, my only problem is that our company has grown to the point where we now have to deal with a committee prior to a major purchase such as this one." K'Narf said, ignoring Toby's flash about what committee. "With that thought in mind, could you provide us with a copy of your video presentation along with the associated costs figures. I promise that we'll get back to you as quickly as possible." He added, sending Toby a flash to back him up and he'd explain later.

"Everything you will need is right here." Joe senior said, taking a small flash card out of his pocket and handing it to K'Narf, disappointment obvious in the tone of his voice but not showing in his face. "As I mentioned earlier though, there is a second party interested in this particular property."

"My father wouldn't try to hustle me, would he?" K'Narf asked, flashing his question as he turned to look at Toby.

"In a heartbeat." Toby flashed back. "Remember, to him it's not personal, it's just good business." He added, struggling to keep a straight face.

"Well, now that our business is out of the way, I do have a personal matter to discuss with you." Joe senior said, turning his attention fully to K'Narf.

"Sir." K'Narf said, caught by surprise.

"It seems that Junior has torn down the old farmhouse and built him and his grandfather a new one and he wants to have a house warming party. Your mother wanted me to ask if you and T.J. would like to come?" Joe senior asked, his eyes lighting up at the mention of T.J.'s name.

"When?" K'Narf asked, still suffering from the surprise of the question.

"Sunday." Joe senior answered, somewhat bluntly. "If you could get there around four there would be time to give your wife a tour of the farm before dinner, I don't believe she's seen it, has she?" He asked, asking his question in an attempt to soften the invitation and make it more appealing.

"Yes, yes, I will of course have to talk to T.J., but I know of no reason at the moment that would keep us from coming." K'narf said, quite pleased with the invitation.

"Good, I will let your mother know." Joe senior said, pushing his chair back and standing up in an obvious act of dismissal. "She will be pleased." He added, as both K'Narf and Toby stood up and shook his hand.

"I will be looking forward to hearing from you." Joe senior said, as he opened his office door for them.

"What committee?" Toby flashed, as he followed K'Narf out into the hall. "That site was exactly what we're looking for!"

"I don't know, something felt wrong! I kept getting the feeling that we needed to hold off, that's all I can tell you right now, something's not right, I can feel it!" K'Narf flashed, not telling Toby that it was the Holy Spirit's presence that was saying no.

Chapter Eight

"That's an awful pretty outfit, I don't think I've seen you wear that one before." K'Narf said, looking over at the dark green pant suit T.J. was wearing as he turned off the highway and started up the drive toward the farm.

"No, I picked it up on Rodeo Drive yesterday." T.J. said, pleased the K'Narf had finally noticed her new outfit.

"Wow, Junior really did build a new house, would you look at that!" K'Narf exclaimed, forgetting all about T.J.'s new outfit as Junior and Gramp's new house came into view; flashing T.J. a picture of what the old farm house had looked like as he spoke.

"Two story, New England facade with gables; yes he did a very nice job. I particularly like the heavy stone he used for the initial six feet of wall." T.J. said, looking at the way the house had been constructed as they drove up and parked, noticing too that they seemed to have arrived before K'Narf's mother and father.

"I don't see dad's car." K'narf said, noticing the same thing as he got out and walked around and opened T.J.'s door.

"Hey K'Narf!" Junior yelled, coming out of the house and heading toward them. The farm life is definitely good for him, K'Narf thought, as he looked at his brother; tanned and at least thirty pounds heavier as he walked up and shook his hand. "Mom said you looked like our great grandfather now, but I thought she was putting me on, I guess she wasn't." He added, as he got a good look at K'Narf.

"Junior, I'd like you to meet my wife T.J. T.J. my big brother, Junior."

"Older brother yes, big brother I don't think so." Junior said, looking up at K'Narf as he took T.J.'s offered hand in very obvious, to the point of embarrassment, admiration. "How did someone as drop dead gorgeous as you wind up with him?" He asked, looking at T.J. as he nodded in K'Narf's direction.

"As you just pointed out little brother, you are...."

"I'll take that as a compliment, but I must warn you that your brother has an offensive streak in him that I cannot always control." T.J. said, cutting K'Narf off as she retrieved her hand. "He's definitely his father's son." She flashed at K'Narf.

"He's going to be a hurt son if he makes anymore cracks like that!" K'Narf flashed back.

"So, what do you think of our new house?" Junior asked, turning his attention back to K'Narf. "Come on in and I'll show you around. Mom's already here with Ada Louise; she sent dad to the store for something they forgot to bring." He added, ignoring his own question as he ushered them toward the front door.

"Mom's already here." K'Narf repeated, understanding now why their car wasn't in the driveway.

"Yes, Mom, Dad and Ada Louise arrived about an hour ago. In fact here comes Dad back now." Junior said, nodding in the direction of the driveway where they could now hear the sounds of an approaching car.

Turning, K'Narf saw his Dad's Lincoln pull up and park behind the Lady, then watched as his father got out and started toward them carrying a brown sack obviously containing whatever it was they had forgotten to bring. "He's hurrying, so let's wait for him." K'Narf said, stopping just short of the front door.

"You made it!" Joe senior said, shifting the brown sack to his left arm so he could take T.J.'s offered hand and with a slight bow give it a very light kiss, totally ignoring K'Narf in the process. "Like I said, Junior is your father's son." T.J. flashed, seeing the look in K'Narf's eyes.

"If they keep this up I'm going to start sticking pins in some very sensitive nerves." K'Narf flashed, referring to what Brent had taught them about offense.

"You'll do nothing of the kind." T.J. flashed back, you'll take their attentions to me as a compliment to your wife.

"Just as long as they don't get too carried away with their compliments." K'Narf flashed back, as they all turned and entered the house.

"K'Narf!" Robbie exclaimed, rushing up and hugging K'Narf, then turning to look at T.J. "Oh my dear, I just love your outfit!" She added, stepping back so she could get a good look at T.J. "Where ever did you find it?"

"A little shop over on Rodeo Drive, picked it up yesterday." T.J. said, comfortable now that she was on more familiar ground; ground where women instinctively understood each other.

"And your necklace, you got that there too?" Robbie asked, getting a closer look at the jeweled necklace hanging around T.J.'s neck.

"No, actually this was a gift to my great, great grandmother from the Emperor back about the time the Colonies were signing the Declaration of Independence. I wore it because I thought the alternating green and red of the starburst cut emeralds and rubies matched the green of my new outfit as well as the dark red embroidering on the sleeves and collar." T.J. said, fingering the finely crafted gold chain of small sparkling rubies and emeralds that was hanging around her neck.

"And that?" Robbie asked, unable to stop herself from asking as she pointed at the large jewel hanging on the end of the necklace.

"That is a very rare Star Safire, it was rare when it was given to my grandmother and it is even more so now." T.J. said, turning it in the light so Robbie could see the beautiful six-pointed blue star with its violet highlights glowing in its center.

"It is gorgeous T.J., your great, great grandmother had very good taste." Robbie said, giving credit where it was due. "It's like it has an inner glow all its own." She added, her eyes suddenly getting wide as she noticed for the first time the inner glow coming from T.J. Herself.

"They'll talk about clothes all night long, come on and I'll show you my new pool table."

Junior said, talking to K'Narf and his father as he broke the connection that had just been established between T.J. and Robbie. "Oh, but first let me introduce you to Cindy, who, as of last night, is my fiancé." He added, as a very pretty girl with big black Italian eyes stepped out of what must have been the kitchen.

"Fiancé! K'Narf exclaimed. "Someone actually agreed to marry you?" He asked, the surprise in his voice as well as the question getting a smile from everyone but Junior.

"K'Narf!" T.J. flashed, scolding him.

"My turn!" K'Narf flashed back.

"Cindy, ignore my little brother and come over here and meet his wife, she'll be your new sister-in-law which will help you get over the shock of K'Narf being your new brother-in-law." Junior said, proving himself quite capable of thinking on his feet.

"What are the rules again about hurting big brothers?" K'Narf flashed at T.J.

"Just one, your not allowed to!" T.J. flashed back.

"You take all the fun out of life." K'Narf flashed, as Cindy came over and shook his hand then quickly joined T.J. and Robbie.

"Come on, I want you to see my new pool table." Junior insisted, trying to get K'Narf to follow him.

"Okay, but first, how much time do we have before dinner?" K'Narf asked, directing his question at his mother.

"About an hour, why?"

"Good, after I go look at Junior's new toy, I want to show T.J. the farm. She's never seen it."

"Toy!" Junior exclaimed, leading the way toward a door at the far end of the living room. "Pool is the game of Kings, just ask anyone." He added, reaching the door and opening it.

"By anyone, you mean a salesman don't you?" K'Narf asked, following Junior and his father into what apparently was Junior's poolroom.

"How far along are you?" Robbie asked, looking at T.J.

"I'm sorry, I uh, don't understand." T.J. stammered, turning to look at her mother-in-law.

"You are radiant my dear, and it all doesn't come from your pretty outfit or that gorgeous necklace so I ask again, how far along are you?" Robbie asked, repeating her question.

"You're pregnant?" Cindy asked, picking up on the conversation.

"No, I mean, I don't know for sure. I've only missed once." T.J. stammered, nervous now that someone else was confirming what she had only suspected up until now.

"That's all it takes dear." Robbie said, pleased with herself that she had recognized that her new daughter-in-law was pregnant. "Does he know?" She asked, indicating K'Narf with a nod of her head at the poolroom door.

"No, and please don't say anything, I want to tell him in my own way and time." T.J. said, pleading with her eyes as well as her words.

"It'll be our little secret, won't it Cindy?" Robbie asked, looking at Cindy in a way that said she had better agree.

"Yes ma'am, I won't say a word." Cindy said, instantly agreeing with her soon to be mother-in-law.

"Okay, here they come, lets go back to the kitchen Cindy and see how Ada Louise is coming with dinner." Robbie said, leaving T.J. as K'Narf, Joe senior and Junior walked up.

"We need to get one of those." K'Narf said, referring to the pool table that Junior had just shown him.

"If that's what you want dear." T.J. said, actually meaning it since she too loved to play pool.

"Ready to go look at my Grandmother's farm, and by the way, where is Gramps?" K'Narf asked, realizing that he had yet to see his grandfather.

"Taking a nap. Mom knew it would be a long evening for him and made him go lie down." Junior said" "And it's my farm now." He added, realizing what K'Narf had just said.

"In my mind it will always be Grandmother's farm, Gramps and you are just inheritors." K'Narf said, taking T.J.'s hand and heading for the front door.

"Do you mind if I go with you?" Joe senior asked, not as sure of himself in a social setting as he was in a business one.

"Of course you can." T.J. said, reaching over and taking his hand, which brought an immediate smile to his face.

"Well, I see this farm six days a week so you go ahead, I'm going to go find Cindy." Junior said, turning and heading for the kitchen as K'Narf, T.J. and his father headed out the front door.

"First I want you to notice that there is a large raised bed built into each corner of the farm. Each bed is three feet high, twenty-five feet long on each leg and twelve feet wide, which actually amounts, when you subtract out the thickness of the block they are built with, to six hundred square feet of viable garden bed for each planter. These are the corner stones that Grandmother built the farm on." K'Narf said, getting a quizzical look from his father.

"Explain please?" T.J. asked, looking for logic in what K'Narf had told her.

"Yes, okay, you see each bed is set up with an automatic watering system, every fifteen days in the summer, and every forty-five days in the winter. We used to use batteries for our timers, but with the advent of solar panels we now use them to power our timers." K'Narf said, continuing his explanation.

"How does that make these planters the cornerstone of the farm?" Joe senior asked, unable to stay out of the conversation any longer.

"That's where Grandmother's genius comes into play." K'Narf said, looking at his father. "Grandmother planted those planters with every possible plant that she knew beneficial insects would like. Alliums, mints, lantana, cosmos, and rubedkia for starters, then she added cilantro, sweet alyssum, California poppies, willows and fothergilla as well as chickweed, wild mustard and dandelions and many, many more that I don't know the names of. The end result was an environment that fostered the breeding and proliferation of thousands of beneficial insects that live and winter over in those beds from year to year. That's the main reasons she never had to use any form of artificial insecticides."

"So that's what those beds are for." Joe senior said, nodding his head in understanding. "I've always wondered, just didn't want to show my ignorance and ask." He admitted, smiling at T.J. and then at K'Narf.

"Then she laid out twenty one, one acre beds as you can see, and built seven quarter acre poly tunnels for growing the farms vegetables. She also said that if the Lord cursed a fig tree because it wasn't producing, she wasn't going to have anything growing on her farm that didn't produce something even if it was just a flower. That's why as you look around you'll notice that every tree and bush on the farm produces something, pecans, walnuts, blackberries, blueberries and even Kumquats." K'narf said, bragging about his Grandmother and how she laid out her farm.

"What about that rotation thing?" Joe senior asked, remembering the many conversations he had overheard on that subject.

"Simple really." K'Narf said, a little surprised at his father's question. "God told the Israelites to rest their land every seventh year, so that's what she did. Every year she took three of the one acre beds and one of the poly tunnels out of service which totaled one seventh of the farm."

"Out of service?" T.J. asked, trying to understand how you took part of a farm out of service.

"Yes, we stripped the poly off and planted cover crops in the greenhouse beds and in the three one acre plots instead of vegetables. As quickly as the cover crops came up we plowed them back into the ground and planted again. We always got five and sometimes six good cover crops plowed back into the ground before the year was up. By the time we were through the ground was a rich black loam and was ready to be planted the next year." K'Narf said, as he swung his arm around at the neatly laid out farm with the seven large poly tunnel greenhouses, one of them being bare metal poles only because of its poly covering having been removed.

"Now it makes sense." Joe senior said, remembering again the many conversations about this very subject, loss of income versus improving the soil and heeding God's word, which he had overheard in the farms old kitchen. "I'm glad she stuck with God's word." He added, now that he understood what it was she had been doing.

"Yes, us too, and we do so love all the vegetables." T.J. said, noticing the sudden change in Joe senior's expression.

"You're the ones buying everything they can produce, aren't you?" Joe senior asked, jumping on T.J. 's comment as he looked at K'narf and then at T.J.

"Yes, but we prefer they didn't know." K'narf said, looking at his father.

"It's not charity, if that's what you're thinking." T.J. said, instantly realizing the secret her words had revealed. "It just so happens that this farm grows the best organic vegetables on the market and we are more than willing to pay somewhat above the going price to get them."

"Yes, that I understand, and no, I will not say anything. I will keep your secret." Joe senior said, not realizing that keeping this simple little secret with his son had created a bond, their first one as father and son. "I think they're ready for us." He added, nodding in the direction of the house where Junior was waving at them to come in.

"Gramps." K'Narf said, seeing his grandfather as they entered the house. "I'd like you to meet my wife T.J." He added, as his grandfather came over and gave him a quick hug without actually looking at him.

"It is my absolute pleasure I am sure!" Gramps said, taking T.J.'s hand and giving it a resounding kiss, then taking a real look at K'Narf he exclaimed, "I'll be a monkey's uncle, you really do look like him. Robbie said you did, but I didn't believe her." He added, moving away from K'Narf like he'd seen a ghost and setting down on the other side of the table before any of the ladies had been seated.

"He's one of the family all right." T.J. flashed at K'narf. "He drooled all over my hand."

"There are finger bowls on the table, so you're saved." K'Narf flashed, noting the beautiful way the table had been set.

"Junior must have bought this just for this occasion." T.J. flashed, noting as K'Narf did the beautiful dining table with its eight elegant dining chairs.

"Mom probably had something to do with it." K'narf flashed back, pulling out T.J.'s chair and holding it for her, then carefully sliding it in as she sat down next to his mother. Setting down next to her he had the uncanny feeling that somehow he was protecting her from the various male members of his own family.

"Now." Robbie said, as Ada Louise came in with her immaculate stainless steel serving cart. "There is no bacon to contend with this time. These are prime Black Angus steaks that have been marinated for just over twenty-four hours. If I hear a single complaint someone gets hurt." She added, as Ada Louise proceeded to serve everyone.

"This is what life should be all about." K'Narf flashed at T.J. "Friends, family and good food." He added, as Ada Louise started serving.

"Yes, I agree whole heartedly!" T.J. flashed back, sending K'narf a picture of a small red heart with the words "I Love You" written on it.

Looking at the sizzling steak that had just been placed on his plate, K'Narf felt the saliva begin to flow in his mouth as he answered his mother. "There will be no complaints, mother, I assure you as long as someone hurries up and says grace before it cools off."

"T.J., would you mind?" Robbie asked, taking T.J.'s hand.

"My phone's vibrating!" K'Narf flashed at T.J.

"Well, answer it, it wouldn't be vibrating unless it was important."

"Now?"

"Yes, now!" T.J. said, letting go of K'Narf's hand.

"Excuse me." K'Narf said, reaching into his pocket and taking out his phone. "Yes, what is it Brent?" He asked, having seen Brent's cell number on his digital readout.

"We need to have a meeting ASAP, and everyone needs to be there!" Brent said, a serious sounding excitement in his voice.

"Right now?" K'Narf asked, wondering what could have happened.

"Yes, as fast as you can get everyone together. It needs to be in the Whale so Rudy and Wanda can be there!" Brent said, again sounding excited and serious all at the same time.

"I've just sat down to dinner with my family, couldn't I call everyone after we've finished. It would only be a couple of extra hours. No one's dead or dying are they?" K'Narf asked, trying to put off what sounded like an emergency for at least a few more hours.

"Is that a question?"

"Yes, I guess it is." K'Narf answered, not liking the sound of Brent's voice.

"Then I have a question for you. When are you going to start taking me serious? You speak of respect, but when I tell you we have a emergency situation and need an immediate meeting what is your response, it's just Brent so let's put it off until it's more convenient." Brent said, obviously angry and getting angrier.

"I am terribly sorry Brent, you are absolutely right and I apologize. We will be there as fast as we can, and so will everyone else. Where are you right now?"

"In South Korea, I can be in the Whale in just under five hours." Brent said, anger fading from his voice.

"So can we, and everyone else should not be far behind." K'Narf said, seeing the screen fade as Brent signed off, wondering why he had used the phone instead of flashing him as he flashed T.J. about what had just transpired between him and Brent.

"He is becoming a man." T.J. flashed back, taking her napkin out of her lap and putting it on the table.

"No! Not again!" Robbie cried, realizing instantly what was happening. "You wouldn't do this to me again?" She asked, leaning over so she could look around T.J. at K'Narf.

"I'm sorry mother, something has come up and I have no choice." K'Narf said, saying "I" so T.J.

wouldn't get any of the blame as he slid his chair back and stood up, salivating as he glanced down at the still sizzling steak. "Could we take our steaks with us?" He asked, timidly looking over at his mother.

"Ada Louise!" Robbie yelled, making sure she was heard out in the kitchen.

"Yes ma'am." Ada Louise said, sticking her head out of the kitchen door.

"Do you have any aluminum foil with you?"

"Yes ma'am."

"Then come out here and wrap up these two steaks, my son and his wife have to leave again!" Robbie ordered, visibly angry. "At least they can take them with them, though I don't know how good they'll taste cold." She added, pushing back her own chair and standing up.

"Moments later they were all gathered outside as K'Narf and T.J., steaks in hand, headed for the Lady.

"Hang on a minute." Robbie yelled. "We've got you blocked in, Joe go move the car so they can get out."

"That's not necessary!" K'Narf yelled back, opening the door for T.J. and helping her in, then running around and opening his own door and jumping in.

"Is this wise?" T.J. asked, just as K'Narf cloaked the Lady and lifted off.

"Brent said to hurry." K'Narf answered, wishing he could have seen his father's face.

Chapter Nine

"Rudy." K'Narf flashed, programming the onboard computer for the Whale as he contacted Rudy.

"Yes, what is it?" Rudy asked, wondering why K'Narf would be contacting him, something he hadn't done much of lately.

"What's the status of the conference room?"

"One group just left, expecting another in three hours and there should be a group loading up right now, why?"

"Can you contact Premo or Sylvia and stop that last group from loading?"

"Let me check, be right back." Rudy flashed, breaking the connection with K'Narf.

"What did he say?" T.J. asked, feeling uncomfortable as K'Narf increased the pressure to a full four G's of thrust.

"He's trying to stop a group that's getting ready to depart for the Whale." K'Narf said, frowning at the time it was taking. He did not like waiting.

"That will be a disappointed group." T.J. said, thinking of the level of anticipation the newbies had for their trip to the whale.

"They aren't the only one's being disappointed today." K'Narf said, thinking of his mother.

"K'Narf?" Rudy flashed.

"Yes, go ahead." K'Narf flashed back, acknowledging Rudy.

"They hadn't left yet. Premo put a hold on them, what's up?"

"Emergency meeting, Brent's on his way and so are T.J. and I. Be there in about five hours, plenty of time to clear out that second group coming your way before we get there. You and Wanda will need to be there too, and tell Redwing I don't want any of her shunning Brent stuff until this is over, this is Brent's show and he doesn't need any more stress than he's already under, okay?" K'Narf asked, flashing his request via Rudy to Redwing as an order rather than asking it as a favor.

"I'll tell her." Rudy flashed. "Is everyone coming, if they are I don't know if we have enough sweet rolls for a group that large, the newbies always bring their own you know. Oh, and you need to plot a hyperbole to the left coming up, returning Limos always go to the right."

"Thanks, I'll do that, but this isn't about sweet rolls, it's about an emergency meeting!" K'Narf flashed back. "Whatever you have will do."

"Okay." Rudy flashed, as K'Narf broke the connection. Must be pretty serious if the boy isn't interested in food, Rudy thought, frowning.

"Well?" T.J. asked, struggling to get comfortable knowing it was going to be a long five hours.

"Premo stopped them so we have the conference room, now I need to get through to

Toby." K'Narf said, mentally dialing up Toby's frequency.

"Tell him to have Irena flash me." T.J. said, beginning to believe she might really be pregnant from the way she was feeling, and praying silently that five hours of four G pressure wouldn't cause her to have a miscarriage if she was.

"Toby, where are you?" K'Narf flashed, the second he got through to Toby's frequency.

"In my office, why?"

"Brent's called an emergency meeting. He's already on his way to the Whale. T.J. and I are ten minutes out and on our way, and we need to get everyone else up there as fast as we can." K'Narf flashed, trying to impress upon Toby that this was truly an emergency.

"Brent's calling an emergency meeting, why?" Toby asked, immediately challenging anything Brent might be doing.

"Because he's the Head of Security, that's why!"

"And if it's not an emergency, what then?" Toby flashed, challenging Brent's authority with his question.

"Then we'll be looking for a new Head of Security, and I can assure you that is not going to happen." K'Narf flashed, knowing Toby's reluctance was based on his competitive reactions to Brent, he added. "A little heads up for you. When you get to the meeting you will not be dealing with Brent the teenager, but with Mister Brent Conrad the man, a proud member of the Nez Perce and direct descendent of Chief Joseph. I'd keep that in mind if I was you."

"He's changed that much?"

"In my opinion, yes. If it makes you feel any better, my first reaction was the same as yours." K'Narf flashed, softening his words by exposing his own initial reluctance to Brent's request.

"What do you want me to do?"

"Contact Premo, Sylvia, Minh Le, Debbie, DeeDee and of course Irena and tell them about the meeting as quickly as possible and tell Minh Le to bring all his recording equipment."

"That will mean more than one van going up, I'll start checking around...."

"No!" K'Narf exclaimed, cutting Toby off. "There are three Limos all prepped and ready to go, tell Premo your going to need one of them. That gives us fourteen seats counting you as the driver. Tell Minh Le to bring Hacker, and contact Karl Grauer and his assistant Mitch Murphy and ask them to come too. That makes ten, might as well come up with a full load so contact Peter and Antoinette Vanlieu and Dan and Susan Ripplinger and ask them to come along as well. That would give you a full Limo."

"Give me an hour and we should be airborne." Toby flashed, fully committing himself.

"Make that a half-hour." K'Narf flashed back, feeling an urgency he hadn't felt before as the Holy Spirit touched him. "Can you sleep?"

"I'll try, but wake me before you flip us over at mid-point, I'll want to brace myself before you do that." T.J. said, closing her eyes and trying to relax, knowing that sleep would be hard to come by under four G's of pressure. Knowing also that K'Narf had forgotten to ask Irena to flash her, but that was okay, she didn't feel like talking to her now anyway.

She still wasn't in a mood to talk when K'Narf asked for the umpteenth time if she wanted her steak. "For the last time I'm telling you that you can have my steak. I don't want it. Just the thought of eating it makes me nauseous." T.J. said, as K'Narf finally proceeded to unwrap her very cold steak, having already eaten his own, and take a bite out of it, making her nauseous just seeing him eat it.

"Thank you." K'Narf said, talking around a mouthful of steak as he watched the Whale grow in his windshield. Cautiously looking around for Brent's van he dove into the Whales eye, set down long enough to match the Whales spin and headed for the airlock. Minutes later he slipped into the Whale proper only to see Brent's red van already sitting on the landing pad. Setting down next to it he quickly got out and went around and opened T.J.'s door, noticing as he did the strained look on her face. "You okay?" He asked, as he took her hand and helped her out.

"Yes, just a little tired. Didn't get much rest or sleep." T.J. said, leaping at the yellow guy rope that led to the kitchen. "Think I'll lay down until the meeting starts." She added, pulling herself slowly toward the kitchen, followed immediately by K'Narf.

"Wanda." K'Narf said, seeing Wanda as they dropped into the kitchen seconds later. "T.J. needs to lie down and rest a little before the meeting starts, can she use your bed, it's closer than ours."

"Yes, of course." Wanda said, immediately stepping over and slipping her arm around T.J.'s waist and guiding her toward her and Rudy's bedroom. "What's the matter?" She asked, as she opened the door and closed it behind them.

"I think I'm pregnant." T.J. said, knowing she could not hide her condition from Wanda.

"Lord help us." Wanda said, literally picking T.J. up and laying her down on the bed like a little doll. "Does K'narf know?"

"No, and I don't want him to know, at least until this meeting is over and we know what has Brent upset." T.J. answered, closing her eyes and fully relaxing for the first time in over five hours.

"You just rest, I'll come and get you before the meeting starts." Wanda said, walking softly over to the door and closing it just as softly behind her.

"Is she all right?" K'Narf asked concerned, since he had never known T.J. when she wasn't up and going with more energy than he had most of the time.

"She'll be fine. I told her I'd wake her when the meeting starts, and Brent is in the Chapel, wants you to call him when everyone gets here." Wanda said, nodding in the general direction of the Chapel. "Oh, and Redwing got your message and said to tell you okay." She added, looking at K'Narf through new eyes herself, he's changing, she thought, not the young boy anymore.

"Where is she, anyway?"

"I sent her over to your bedroom to feed Thomas and Judy, didn't want them causing a commotion during the meeting."

"And Rudy?"

"Messing with his garden, where did you think he'd be?" Wanda said, grinning.

"Good, you wouldn't have a cup of coffee around here somewhere would you?" K'Narf asked, glancing at the conference room door; feeling as though it was calling him.

"You go on in and find yourself a seat, I'll bring a cup and a sweet roll right in, you take both cream and sugar, don't you?" Wanda asked, as she turned and stepped softly over to her small but neat kitchen area.

"Yes, thank you." K'Narf answered, heading for the conference room and looking around. It seems so peaceful, he thought, looking at the softly lit room with its big polished conference table as he stepped over and pulled out the first chair on the opposite side of the table. He wanted to be facing everyone as they entered, purposefully pulling the chair next to him closer as a sign to everyone coming in that it was reserved for T.J.

"Here you go hon." Wanda said, walking in and setting down a tray and carefully placing a steaming cup of black coffee in front of K'Narf along with a soy creamer, sugar bowl, a small plate with a large freshly baked cinnamon roll on it and a napkin wrapped around a fork and spoon. "That should hold you until everyone gets here." She added, as she turned and left, leaving K'Narf to his private thoughts again.

"Oh Lord!" K'Narf said, bowing his head. "I don't know what has put Brent in such a state of panic, and because I don't know I have no idea what to ask You for. All I can do is ask that You approve and bless our decisions and that at the end of the day it's Your will that we will be doing. Oh, and please bless my wife, it's not like her to be tired like this, and please bless this food to my body." He added, as he raised his head and unwrapped his silverware.

"K'Narf, Wanda said you were in here." Rudy said, stepping into the room and pulling out a chair on the opposite side of the table about three chairs down from the front. "You'll have to excuse my gardening clothes, managed to wash up but there was a little angel sleeping in the bedroom and Wanda wouldn't let me go in and change." He added, watching as K'Narf struggled to chew and swallow a large bite of cinnamon roll.

"Rudy!" K'Narf exclaimed as he managed finally to swallow, surprised by Rudy's sudden entrance as well as his appearance. "Man have you ever changed." He added, giving Rudy a thorough once over with his eyes as he took a swallow of coffee.

"Yeah, I look about the way I looked when I joined the Navy back in the thirties." Rudy said, grinning.

"How come you didn't get rid of those scars?" K'Narf asked, taking another bite as he looked at Rudy's hands which still bore some serious scars on his knuckles.

"Earned them scars the hard way, didn't want to get rid of 'em." Rudy said, holding up his hands and examining them.

"Earned them, how?" K'Narf asked, talking around another mouth full of cinnamon roll.

"Fighting in smokers."

"Smokers?"

"Yeah, smokers. You get several hundred men below decks lounging around a trumped up boxing ring on a Friday night with the smoking lamp lit, and pretty soon the air was so thick with smoke you could barely see so we called them smokers." Rudy said, trying to describe the way things used to be in the Navy he remembered.

"What did you fight about?" K'Narf asked, trying to understand the culture and times Rudy was talking about.

"Just about anything, didn't need much of a reason. If someone was giving you a hard time and you thought you could lick him, you'd challenge him to a fight. If he refused he was a chicken." Rudy said, grinning at the memories he was dredging up.

"And you got into a lot of fights?"

"I was a cook and I was black, but in the ring there was no rank or race just bare knuckles, so yeah, I got into a lot of fights. Made a lot of money too, at least at first." Rudy said, his grin getting bigger.

"How'd you do that?" K'Narf asked, trying to get the last dribbles of frosting off the plate by scrapping at it with his last bite.

"Had my buddies place bets on me. Worked pretty good until they figured out that I could just about whip the tar out of anyone foolish enough to get in the ring with me."

"And that's how you got those scars?' K'Narf asked, pointing at Rudy's hands with his fork.

"Yeah, see this one right here." Rudy asked, pointing at a nasty looking scar between his first and second knuckle on his right hand. "Big boatswain's mate thought he could whip me, so he referred to me with a racist slur in order to get me to fight him. Well he got his fight and I hit him so hard I imbedded one of his eyeteeth right here between my knuckles. Corpsman had to pull it out with a pair of tongs, it hurt like the very devil when he did it too."

"What happened to the boatswain's mate?"

"Called him snag after that." Rudy said, grinning at a memory he hadn't thought about for more years than he could count.

"Snag?"

"Yeah, short for snaggle tooth." Rudy said, still grinning.

"I see." K'Narf said, his curiosity satisfied about Rudy's scars. "How's that garden of yours coming?" He asked, changing the subject as he laid his fork down on the empty plate and picked up his coffee cup; pushing the plate and fork into the center of the table at the same time.

"Great, you need to take a few minutes and come down and look at it when you get a chance." Rudy said, obviously excited about his garden.

"I noticed you've added a lot of lights down there." K'Narf said, thinking how it had looked like a lit up football field the last time he had looked at it.

"Got an area two hundred by a hundred and fifty feet under lights, with a soil depth of just over three feet." Rudy bragged. "Placed the lights twelve feet up with enough lumens to light up every square inch of it too.

"Plants growing up here okay?" K'Narf asked, taking a sip of coffee as he noticed the sparkle in Rudy's eyes as he talked about his garden.

"Lot of trial and error." Rudy said, his brow wrinkling in thought. "Been planting heirloom seeds of just about everything. Finding quite a few of them that seem to do real good up here too."

"Heirloom?" K'Narf asked, remembering that his grandmother used the same kind of seeds.

"Yeah, if I find something that grows up here, I want to be able to harvest seeds and grow it again and you can't do that if you'd grown it from a genetically modified seed." Rudy said, thinking K'Narf should have known that having been raised on a farm.

"What kind of things, I mean the ones you said are doing real good up here?" K'Narf asked, thinking he really did need to go down and look at Rudy's garden the first chance he got.

"Well, planted over twenty different tomatoes and found three that are doing real well. Taste good too, and a cucumber that is small enough to make pickles out of and yet it still tastes good raw. Big stuff seems to get all puffy inside like watermelons and cantaloupe, but small ones seem to do okay."

"Small ones?"

"Yeah, like I found a small watermelon that was about six inches in diameter, and a small cantaloupe about three inches in diameter that grow really well up here. Iceberg lettuce didn't do well, but almost all of the leaf lettuces do and so do the onions and garlic, but I'm mostly proud of my trees."

"Trees?" K'Narf asked, surprised by Rudy's mention of trees.

"Yeah, knew we'd be needing something to give us some natural oxygen, so I planted a whole bunch of different trees. Couldn't find one that could take the light and gravity until one of the newbies brought me up one of those little ornamental fern trees, the kind you find around Christmas."

"Yeah, I know the ones. Junior bought grandmother one and she didn't like it because it didn't produce anything." K'Narf said, remembering how Junior had been hurt by her refusal to keep it.

"Everything God created has a reason, in this case it was oxygen. I have over thirty of those little trees now and some of them are over eight feet tall. I've been keeping a real close watch on the amount of oxygen we use and it's gone down enough to be measurable since I've planted those trees. I think if I could get about two hundred of them growing real good we wouldn't need to depend on a fresh oxygen tank being hauled up here every couple of months." Rudy said, seeing that K'Narf was both surprised and pleased about his trees.

"You've become a real master gardener, haven't you?"

"Don't like the word 'master', if you want to call me something just call me the Head Gardener." Rudy said, having been thinking of himself that way for some time now.

"Head Gardener?"

"Yeah, Brents the Head of Security, DeeDee's the Head of Procurement for food and stuff, and Toby is the Head Chief of Staff or something like that, so I figured I'd start calling myself the Head Gardener." Rudy said, grinning at K'Narf, realizing as he said it that he was only half kidding.

"And what am I?" K'Narf asked, trying to trip Rudy up.

"You, you're just the Head, you ain't got no other title." Rudy said, his grin getting even bigger if that was possible. "Uh oh, they're here. Wanda just flashed me that the Limo just pulled in and that they'll be here in just a few minutes. Let me get rid of that plate for you." He added, picking up K'Narf's dirty plate and heading for the door.

"Brent." K'Narf flashed. "The limo just pulled in, we're all here." He added, knowing Brent had heard him. About time to find out what this is all about, K'Narf thought as Rudy returned and sat back down.

Chapter Ten

"Well, we're here." Toby said, as he walked into the conference room followed by Minh Le who immediately began setting up his recording equipment with its ubiquitous digital recording globe. "Where's he at?" He added, looking around the room.

"He's still in the Chapel, he'll be here in a minute." K'Narf said, knowing Toby meant Brent. "Why don't you grab those two front chairs for you and Irena." He added, pointing at the two chairs opposite his and T.J's. Then realizing that he hadn't set one up for Brent he quickly got up and pushed the chair he had been sitting in over to the front of the table, pulled the one he had been saving for T.J. over for himself and grabbed the next one down the table and pulled it up next to him for T.J. Barely making it before the room started filling up with silent slightly nervous people. Almost not recognizing DeeDee now that she had a full head of dark auburn red hair and no wrinkles.

"Ain't enough chairs, someone's going to have to set on that second level." Toby said, helping Irena to her chair.

"Not a problem." Dan said, as he took Susan's hand, stepped up to the second level and headed to the back of the room and sat down, followed immediately by Peter, his wife Antoinette and Hacker leaving the chairs around the conference table for everyone else.

They're a tight knit bunch, K'Narf thought, noticing that all the volunteers for the second level were pilots.

"Who wants coffee and a sweet roll?" Wanda asked, coming into the room with a tray loaded down with two full coffee carafes, soy creamer, sugar and multiple stacks of mugs, followed immediately by Redwing with another tray loaded down with cinnamon rolls and plates, forks and spoons. "Start passing them around while I go and get the rest." Wanda said, as she and Redwing set their trays down on the table.

"T.J." K'Narf said, jumping up as T.J. entered the room carrying another tray loaded down with more cinnamon rolls, plates and forks.

"You just set yourself down girl, we can handle this!" Wanda scolded, as she took the tray out of T.J.'s hand and set it on the table.

"Here." K'Narf said, pushing his chair back and pulling T.J.'s chair out for her. "You feeling better?" He asked, as he slid her chair up to the table and sat back down himself.

"Yes, I am." T.J. said, smiling at K'Narf as she patted his hand. "That coffee sure smells good." She added, reaching for one of the cups.

"Redwing, why don't you help the back row with their coffee and sweet rolls?" Wanda asked, noticing that the pilots were unable to help themselves from the trays being pushed down the table.

" My pleasure." Redwing said, moving quickly and gracefully to the back of the room and serving both coffee and cinnamon rolls to the silent but obviously grateful pilots.

"Thank you." Susan said, echoed immediately by Hacker, Dan, Peter and Antoinette.

"Okay." Wanda said, looking around the room and seeing that everyone had been served, then she pulled out her chair and sat down next to Rudy.

"Redwing, where are you going?" K'Narf asked, noticing that Redwing was heading for the exit tunnel.

"I was going to clean up the kitchen, why?"

"No, you're part of this now, I'd like you to stay." K'Narf said, pointing at an empty chair between Wanda and DeeDee.

"Yes sir." Redwing said, embarrassed that everyone was looking at her.

"I think you intimidate her." T.J. flashed.

"I don't mean too, and what's keeping Brent?" K'Narf asked, flashing T.J. as he took a sip of his coffee.

"Ask him, he just walked in." T.J. flashed back.

"Sorry I'm late, I've been getting some wild messages from one of my pick-up teams. It seems that there has been a severe earthquake just off Japan." Brent said, looking around the room as it registered on him as to who was there, including Redwing as their eyes locked for an embarrassed second.

"How bad?" Toby asked, knowing that Japan was one of his key suppliers.

"Real bad, they're being hit right now with a tsunami estimated to be over thirty feet." Brent said, nodding at his row of pilots in the back of the room, wondering at the same time why K'Narf had invited them, but grateful that he had.

"Do we have suppliers in Japan?" K'Narf asked, looking over at Toby.

"Yes, they make the entire firing console for the gunnery cockpit."

"Is your supplier factory anywhere near the Pacific shoreline?" Brent asked, looking down at the seated Toby.

"Yes, I'm afraid it is."

"Are they your sole source?" K'Narf asked, knowing by the look in Toby's eyes that they were.

"Yes, I haven't been able to set up a second source yet."

"How about the people, did they get any warning?" T.J. asked, her immediate concern being friends of hers that lived in Japan.

"Don't know, but a thirty foot plus high tsunami would be devastating even with a warning." Brent said, seeing instant compassion and concern in T.J.'s eyes.

"Sylvia, isn't one of our current newbies from Japan?" Premo asked. "See if you can flash her, maybe she knows what's happening." He added, causing everyone to turn and look at Sylvia.

"No!" K'Narf said, drawing everyone's attention back to himself as he turned to look at Brent. "That is not why this meeting was called, is it?"

"No, no it's not." Brent said, instantly getting everyone's undivided attention. "We have a disaster of our own to contend with." He added, calmly making eye contact with each person as he looked around the room, including Redwing.

"Go on." K'Narf said, confirming in everyone's mind that this was Brent's meeting.

"K'Narf, when we were last together you loaned me your Lady because of her new Cloaking paint, and I headed out to Pluto to recover what I believed was a downed probe. Dan and Susan." Brent said, stopping to point out Dan and Susan sitting in the back row. "Met me out at the Snark's mine site where one of my preset lasers had brought the thing down and helped me roll it up a makeshift ramp and into your van's loading bay. From there I flew it to my compound in South Korea and with the help of my guys took it apart.

What we found was in fact a highly sophisticated probe with a positronic brain that had been severely damaged by an extremely lucky, no, make that blessed, laser hit. That probe was as close to being indestructible as highly skilled engineering could make it, and yet a single hit from a quarter inch laser brought it down. A one-in-a-million hit if you want to think of it in the context of mathematical odds.

What we found inside is the reason I asked you to call this meeting." Brent said, looking at K'Narf. "That probe, and the two still running around were programmed to send a monthly update missile back to their parent fleet; an incoming parent fleet and it only had thirteen missiles left." He added, looking at K'Narf, then looking around the room as what he had just said began to sink in.

"Wait a minute, your telling us that we only have thirteen months before that Snark fleet will be on our doorstep?" K'Narf asked, lunging forward in his chair.

"Yes, that's exactly what I'm telling you." Brent said, confirming for everyone in no uncertain terms, what the situation was. "And with that said, I think this meeting now belongs to you." He added, looking at K'Narf as he scooted his chair off to the side and sat down next to Toby.

"I, uh." K'Narf stammered, as he stood up, accepting Brent's relinquishing of the meeting by the unconscious reflex of standing up. Everyone was trying to talk at the same time, until one by one they fell silent as they turned to look at him.

"K'Narf, take it by the numbers!" T.J. flashed, trying to help.

"Yes, by the numbers." K'Narf said, softly as he shook his head as though clearing out unwanted thoughts, and then looked solemnly around the room until his eyes rested on Karl. "Karl, how many fighters have we completed to date?" He asked, turning everyone's attention to Karl.

"Seventeen." Karl answered, having that number in the forefront of his mind.

"What's your rate of production and how many more do you have parts on hand to complete?"

"By dividing each day into four shifts and running six days a week we are rolling a finished fighter off the line every seventy two hours and we have enough parts for four more of them." Karl said, looking at Toby to see if he agreed.

"Toby?" K'Narf asked, knowing Toby had been acting as central purchasing for the assembly line.

"A shipment of twenty five gunnery consoles were supposed to have left the factory this morning, they should have been at a Tokyo transfer station by now if they shipped on time. All other key parts are either on hand, on order or easily available." Toby said, silently praying along with everyone else in the room that the critical consoles had shipped on time.

"How many people do you have helping you on parts purchasing?"

"Three, why?" Toby asked, not sure where K'Narf was going with his questions.

"Karl, how many parts are involved in building a completed fighter?"

"Counting missiles?"

"Yes, a fully armed and functional fighter including any special equipment needed by the pilots."

"Give me a minute." Karl said, frowning as he looked down at the table and began adding up parts lists. "One thousand and seventy two counting nuts and bolts." He finally said, looking up at K'Narf as he completed his mental listing.

"Premo." K'Narf said, looking down the table at Premo. "I want you to check the records of all the newbies that have come through your school and look for anyone that has had experience in purchasing, particularly if their experience was in a large company and send them to Toby for an interview. Toby I want you to select one of them to take over as your manager of assembly line purchasing and make sure that every critical part has a specific purchasing agent assigned to it."

"Yes sir." Toby said, knowing that the urgency of the situation had changed, but still feeling as if he had just been chastised.

"Brent." K'Narf said, turning his and everyone else's attention to Brent. "How long would it take you and your pilots to get that mother ship functioning and into space, including tacking on the severed front control section and taking that downed scout ship along with her?"

"Why?" Brent asked, caught off guard by K'Narf's question.

"Two reasons. First, you told me that all of that equipment in your asteroid could be transferred to the Whale where we could use it in a jury-rigged command center. Instead I want you and your pilots to totally refurbish that ore freighter into a state-of-the-art command ship. But first how long will it take you to get her back into space?"

"State-of-the-art?" Brent asked, again surprised by K'Narf's question.

"Yes, stealth technology, Cloaking paint, the works, but first how long would it take you to get her back into space?" K'Narf asked again, pushing Brent with his question of how long it would take just to get her off the ground.

"Two weeks." Brent said, excitement tingeing his voice. "We should be able to get her back into space in two weeks." He repeated, glancing down the table and up at his pilots on the second level as they all nodded their heads in agreement.

"Good, Premo, how many of us are there now, and how many newbies are you processing on a weekly basis?" K'Narf asked, as everyone now turned their attention to Premo.

"Right now we number sixteen thousand and forty four, and with nine Limos running newbies up here for orientation six days a week we're adding seven hundred and thirty eight to that number every week." Premo said, knowing his population numbers by heart.

"Minh Le, do you see any problem maintaining that number?" K'Narf asked, knowing Minh Le was coordinating the pick-up teams.

"No, the Lord seems to be pacing our finding them to our ability to train and process them." Minh Le said, confirming what Premo and Sylvia had known for weeks now.

"How many pilot teams have you identified and accepted?"

"Seventy three counting the two teams Brent accepted last week, and I have seven volunteer teams waiting to be interviewed right now." Minh Le said, glancing over at Brent.

"Brent, I want you select and train three hundred teams and you will need to change your training from tandem fighter teams to single fighter training, lone wolf training so to speak." K'Narf said, looking back at Brent.

"Why?" Brent asked, knowing the answer to his own question as soon as he had asked it.

"At our current production rate we will only have one hundred and twenty two fighters going up against nine hundred and sixty and with those odds we won't have the luxury of fighting in teams."

"What about our second factory?" Toby asked, thinking of the factory Joe senior had found for them.

"Yes, we are going to need a second factory, it will help reduce those odds but it won't be the one my father found for us." K'Narf said, seeing the confusion in Toby's eyes. "It's getting far to dangerous to have both factories at risk of being damaged or destroyed by natural disasters like an earthquake or tornado, not to mention overflowing rivers or blizzards stopping production."

"You said you had two reasons for wanting me to get that mother ship off Pluto, what was the second reason?" Brent asked, thinking he knew the answer before he asked the question.

"Mitch Murphy, can you run the assembly floor if I commandeer your boss?" K'Narf asked, nodding at Brent in a way that let Brent know that he had guessed right.

"Yes sir." Mitch said, glancing at Karl to see if he agreed with him.

"Good, Karl, if you were given carte blanche to go out to Pluto and build a dome over an open pit mine." K'Narf said, flashing everyone a picture of the Snarks mine as he talked. "And put an atmosphere in it capable of supporting a work force large enough to develop and start producing a second line of fighters; a line that wouldn't be susceptible to anything going on down here on Earth. Could you do it?"

"That's a big mine, what's the distance across that circular center area, the base area, the area that Snark ship is setting on?" Karl asked, making sure everyone knew what area he was talking about.

"You're looking at approximately nine hundred feet across the diameter." Brent said, answering the question for K'Narf.

"And each of the levels?

"Fifty feet up and roughly one hundred yards deep, with all six levels being almost identical in height and depth." Brent said, again answering for K'Narf.

"And you want the whole thing covered?" Karl asked, making sure he fully understood what was being asked of him.

"Yes, if you don't cover the whole thing you wouldn't be able to camouflage it."

"Camouflage it?"

"Yes, when you're done it should look like a normal part of the planet, at least from space, and any Snark ship or probe flying over it shouldn't be able to tell it's there." K'Narf explained again, trying to give Karl and everyone else an idea of what he was talking about as he flashed them a picture of the mine location covered over with rocks and dirt so it looked like the surrounding terrain.

"Sweet Jesus, you don't ask anything easy do you?" Karl asked, looking up at K'Narf.

"If it was easy, I'd do it myself." K'Narf answered, the look on his face telling everyone that he actually meant what he said.

"Carte Blanche?" Karl asked, frowning in thought as he asked the question.

"Yes, and I mean Carte Blanche. Everyone who has not been assigned to a specific job would be at your disposal, which from what we just heard from Premo will number well into the thousands, and we will make sure that whatever monies you may need will be available to you through the general fund." K'Narf added, seeing T.J. nod her approval out of the corner of his eye.

"What about Snark suits for my crew?" Karl asked, as he in effect said yes to K'Narf's original question as to whether or not he could do it.

"Minh Le?" K'Narf asked, knowing that Minh Le had been growing suits for all the newly accepted pilots. "What's the process for growing Snark space suits?"

"Simple really, once we figured out that the suits were made of a living material it was just a matter of figuring out how to grow them on a mannequin."

"Living material?"

"Yes." Minh Le said, realizing that more than one person had asked the question. "When they started to turn grey and die it became apparent rather quickly that they were actually composed of a living organism. Feeding them when they were not being worn turned out to be rather simple, and cutting one up for seed allowed us to start growing custom fitted suits of our own."

"Custom fitted?" The question again coming from more that one voice.

"Yes, I have the computer scan the recipient's body and then create a headless mannequin to match, install a helmet ring and hermetic zipper seal then turn it upside down, coat the mannequin with a growing medium and place some starter material on each heel. Takes about a week to grow a completed suit."

"How many helmets, helmet rings and hermetic zipper seals do you have in stock?" K'Narf asked, wondering how long it would take to get enough suits made for Karl's initial workforce.

"I was told we we're going to train two thousand pilot teams, so I ordered six thousand sets figuring other people would start wanting suits besides the pilots." Minh Le said, seeing the surprise in K'Narf's eyes when he mentioned six thousand sets.

"You have enough helmets, and helmet rings and hermetic zipper seals to grow six thousand suits?" K'narf asked, repeating everything Minh Le had said to make sure he'd heard right.

"Yes, less the one hundred and fifty seven I've already made."

"Karl, get over to Minh Le's as soon as we get back and get yourself scanned for a mannequin and start picking your men and Minh Le, order up material for at least twenty thousand more suits on top of what you already have." K'Narf said, feeling better as things began to fall into place in his mind. "And DeeDee, you designed that tractor-trailer rig of yours for space, how about showing Karl what you've done. He's going to need quite a few of the big rigs himself before this is over." He added, looking over at DeeDee.

"It would be my pleasure." DeeDee said, smiling over at Karl.

"And don't be using anymore emergency room images for punishment when someone upsets you." K'Narf added, suddenly remembering his promise to David Ben Ryan. Noticing that Karl, Mitch Murphy and the pilots were the only ones that seemed to know what he was talking about as they nodded their heads in agreement.

"Okay." DeeDee said, caught off guard by K'Narf's request. "But you're taking all the fun out of life." She added, complying but complaining at the same time; wondering how K'Narf had found out about her way of punishing her swampers if they messed up.

"Fun for you maybe!" Karl exclaimed, giving DeeDee a raised eyebrow look.

"Could I make one exception to that little request of yours Mister K'narf?" DeeDee asked, glaring back at Karl.

"No, and the two of you play nice, we don't have time for any personal gotcha games." K'Narf said, looking at both of them and wondering if had been a good idea of putting the two of them together, even if it was for just a little information swapping. "And DeeDee, they're going to need to be fed up there. How about setting up a commissary where they can come in and get something hot to eat or just take a break?" K'Narf asked, throwing his impromptu question at DeeDee.

"Commissary, what do you mean commissary?" DeeDee asked, thinking she already had enough to do.

"Don't get excited, you wouldn't have to run it yourself, just set it up and keep it supplied with food. Something with a small airlock so they could come in and get warm while they took a break and ate a little something." You could start out with a kind of gee dunk for hot coffee and sweet rolls until you could get a real kitchen going." K'Narf said, picturing something like the small gee dunk they had set up on the second floor of the factory where he went to get his morning coffee and roll.

"And I wouldn't have to run it myself?" DeeDee asked, making sure she had heard right.

"No, like I said, you'd only be responsible to set it up and keep it supplied with food, someone else can run it for you."

"How about you?" DeeDee said, elbowing Redwing. "How would you like to run my commissary for me?" She asked, surprising Redwing along with everyone else in the room with her offer.

"I uh, I don't know, I'm working here with Wanda and Rudy." Redwing stammered, looking over at Wanda.

"That's too exciting to turn down, you go for it girl, Rudy and I can make out." Wanda said, smiling at Redwing as she elbowed Rudy to stay out of it before he said something.

"What about it?" DeeDee asked, seeing Redwing begin to waver.

"Well, I guess so. If you're sure you want me." Redwing said, seeing a smile light up Karl's face.

"I'll be her first customer, you can bet on that!" Karl said, grinning at Redwing as he said it.

"Not if I get there first!" Hacker said, surprising everyone, including himself, by speaking up.

"And don't worry about a building, me and my crew will see to that." Karl said, turning to look at DeeDee as he said it, noticing that she seemed to be angry about his taking over the building of her commissary, at least that's what he thought she was angry about.

"Grass should not be allowed to grow under one's feet Little Eagle." T.J. flashed at Brent, having noticed the surprise and then irritation that flashed across Brent's face.

"What can I do grandmother, her anger doesn't allow her to speak to me." Brent flashed back, looking over at T.J.

"Then you must speak to her. Did you think her pretty in the princess outfit she wore for you?"

"Yes, she was beautiful!" Brent flashed, grinning as he gave T.J. a slight, hardly noticeable, bow of respect.

"Then when you apologize, you must tell her that." T.J. flashed, answering Brents bow of respect with one of her own.

"Apologize, for what?"

"In your case for breathing." T.J. flashed, smiling across the table at him. "Why don't you try apologizing for tying her hair to that chair for starters?" She asked, getting serious. "That would be a good start."

"You really think that would...."

"Okay, if that's settled, I guess we have a volunteer to run DeeDee's commissary as soon as Karl can get it set up." K'narf said, unintentionally interrupting the silent conversation that had been taking place between T.J. and Brent. "And Minh Le, I appreciate your editing out that kitten attack during our last meeting, but I think every word spoken here today should be put into an emergency reader that needs be distributed and read as quickly as possible so no editing, okay?"

"Yes sir." Minh Le said, having been thinking the same thing himself.

"Is there anyone unsure or confused about what needs to be done?" K'Narf asked, calmly looking around the room and making eye contact with everyone as he did so. "Good, then if you'd all bow your head for a moment of silent prayer, we will adjourn this meeting." He added, knowing now, as he bowed his head, why the Holy Spirit had told him not to buy that second factory as he thanked the Lord for showing him a plan of action that seems to have been accepted by all concerned, which is a miracle in itself, he thought, as he thanked the Lord a second time.

"This meeting is adjourned." K'narf said, a few minutes later when the last head in the room had raised and all eyes were looking at him. All I have to do now is tell my father that I don't want his factory after all, he thought, stepping over and helping T.J. up.

Chapter Eleven

"Toby." K'Narf said, looking up and seeing Toby standing in the doorway. "I'm glad you're here, maybe you can explain these purchasing figures for me." K'Narf said, staring at the screen of his laptop.

"What purchasing figures?" Toby asked, walking around K'Narf's desk and looking over his shoulder thinking K'Narf was talking about purchasing figures for the Tear Drop Fighters.

"These." K'Narf said, pointing at a long list of figures on his screen. "What in the world are they building out there?" He asked, pointing at the dollar total listed at the bottom.

"Four hundred and seventeen million, three hundred and ninety seven thousand and forty four dollars and eleven cents." Toby said, reading the whole number out loud. "You should never have told Karl he had Carte Blanche." He added, uttering a long slow whistle as he said it.

"I know, wait until T.J. gets a peek at this." K'Narf said, scrolling back up to the top of the list.

"What in the world are they buying?" Toby asked, as K'Narf reached the top of the list and stopped scrolling. "I mean they've only been at it what, six weeks?"

"Yes, it's only been six weeks and three days since we had our meeting, and what they're buying is what I'm trying to figure out." K'Narf said, as he pointed at one of the line items. "Seventy two tons of pea gravel in twenty-four forklift carry bags at sixty dollars each."

"No, no, look here, they got a credit of fifteen dollars each for bringing the bags back." Toby said, seeing the credit for three hundred and sixty dollars for twenty-four heavy duty carry bags.

"Yeah, but that's peanuts, look at this line item." K'Narf said, pointing at a listing that had a cost figure of seven hundred and eighty eight thousand, five hundred and forty four dollars and all it said was seventy two 'C' bases. "I'd sure like to know what they're buying?"

"Your guess is as good as mine." Toby said, as K'Narf dragged his cursor down the list.

"Here's another big one."

"Five hundred and twenty three thousand and eighty dollars for seventy two 'C' tops." Toby said, reading the complete line item.

"Uh, excuse me." DeeDee said, sticking her head into K'Narf's office looking rather pretty wearing a bright red car coat over her bluish green space suit with it's glistening silver helmet ring around her neck like a fancy solid ring necklace.

"DeeDee!" K'Narf exclaimed. "Glad you're here. Maybe you can explain some of this stuff to us." K'Narf said, motioning to DeeDee to come into the room and join him and Toby. "What are 'C' tops and 'C' bases?" He added, pointing at those two line items with his finger.

"How many?" DeeDee asked, looking at what K'Narf was pointing at.

"Seventy two of each of them." Toby said, getting into the conversation.

"Seventy two, that's got'ta be columns, there's seventy two of them."

"Columns, that has to be what the 'C' stands for, it stands for columns." K'Narf and Toby said in unison.

"Have either of you been up there yet?"

"No, but looking at these figures I think we need to, why?" K'Narf asked, speaking for both of them.

"You seem to be sorta out of the loop, so to speak." DeeDee said, wondering if she should say anything or not.

"What do you mean?" K'Narf asked, as he spun his chair around to face the two of them.

"Well." DeeDee said, taking a deep breath. "You wouldn't be asking questions like this if you knew what was going on up there."

"I see what you mean." K'Narf said, looking uneasily at Toby. "Has Karl got the dome on yet?" He added, thinking it a reasonable question as he looked back at DeeDee.

"Oh heavens yes! Not only is it on, but it's covered up, and we don't need to wear these anymore since they got the new force fields in place and gave us a breathable atmosphere." DeeDee said, pulling at her space suit as she talked.

"Force fields?"

"Yeah, they copied them from that Snark ship, don't need airlocks anymore now. Kinda neat, but they tingle when you go through them though." DeeDee said, wiggling a little bit as she talked about going through a force field.

"But how, I mean it hasn't been that long?" Toby asked, feeling a little stunned by what DeeDee was telling them.

"You ever watch that Extreme Makeover show on T.V.?" DeeDee asked, looking at K'Narf and then at Toby.

"Yes." Toby said, knowing K'Narf wouldn't have seen it. "They build a home from the ground up, including landscaping and furniture in one week. I wouldn't have believed it if I hadn't seen it myself, why?"

"They got a hundred times more manpower than that makeover show ever thought of having, and what them boys can do in just one shift's enough to make your eyes pop out." DeeDee said, grinning as she bragged on Karl and his men.

"Is Brent up there?" Toby asked, looking more at K'Narf than at DeeDee.

"Yes, he and his pilots are working around the clock on that new command ship of theirs. They used to all come down to the commissary to eat." DeeDee said, thinking how rowdy they were when they all showed up at the same time.

"They eat at your little commissary too?" K'Narf asked, still thinking of a small gee dunk type of operation.

"Well, not any more, they have their own galley where they eat most of the time now, but they're always welcome." DeeDee said, surprised by the question, a fact that K'Narf did not miss.

"Exactly how big is your commissary?"

"Well, it only seats three hundred at a time, but we keep it open around the clock. The steam line's almost a hundred feet long all by itself." DeeDee said, finding it hard to believe that K'narf and Toby didn't seem to know what was going on out at the mine site.

"Steam line?"

"Yeah, it was the only way we could feed that many shifts. Everything you can think of, hot gravies, all different types of potatoes, hot and cold vegetable dishes including salads, dozens of meat dishes, breads, fruit, olives, pickles, and more sweet rolls and deserts than I can count plus coffee, tea, fruit drinks and of course water and it's all available twenty-four hours around the clock. Why?" DeeDee asked, having a serious problem believing the K'Narf and Toby were ignorant of her new commissary too.

"How many cooks?"

"I think Redwing has over a hundred now, all volunteers, newbies that used to cook or just loved doing it. Karl built the commissary with two stories. Redwing and I have offices down on the first level in the kitchen along with the Head Shift Cooks, but we bunk upstairs like everyone else when we're off duty."

"How many people are up there now?" Toby asked, beginning to realize just how far "out of the loop", as DeeDee had put it, that they were.

"Well, we're feeding just under ten thousand a day now. Karl's already planning an addition on the dining area, though."

"Where do they all stay?" K'Narf asked, trying to picture the bedlam DeeDee was describing.

"Well those that don't work in the commissary mostly stay in tents."

"Tents!" Toby said, thinking of the small army tent he used to use in the field.

"Yeah, every size, shape and color you can imagine scattered all over the place, why?"

"You thinking what I'm thinking?" K'Narf asked, looking at Toby.

"Yeah, you know what he's going to say." Toby said, knowing K'Narf would know that he was talking about Brent. "I think we'd better get ourselves up there as fast as we can."

"I think you're right." K'Narf said, looking at DeeDee. "Oh, what was it you wanted to see me about?" He asked, remembering that DeeDee had wanted something when she came in.

"Well, I'm getting ready to run a load up there, but first I need to stop by my warehouse and pick up the rest of the spices so the cooks will stop yelling at me and I need a swamper to help me load'em." DeeDee said, frowning at K'Narf and Toby. "But everyone here's got an assigned shift and Mitch won't let me have anyone and I thought you might be able to help me find someone.

"Could you manage two swampers in that rig of yours?" K'Narf asked, seeing Toby nod in agreement.

"Yeah, if one of 'em doesn't mind riding up in the sleeper, why?"

"You just got yourself two swampers, how soon do you want to leave?" K'Narf asked, pointing at himself and Toby as he said two swampers.

"Do you have suits?" DeeDee asked, wondering if K'Narf and Toby had ever had one made.

"Yes, haven't worn it yet, but I've been keeping it fed." K'Narf said, looking over at Toby.

"Yeah, me too. Just fed mine this morning and I haven't worn mine either." Toby said, seeing a grin flicker across DeeDee's face. "Why?" He added, wondering what had made DeeDee grin like that.

"Nothing, but it's gon'na be interesting to see your faces the first time you have a B.M. in one of these things." DeeDee said, grinning again as she tugged at her space suit. "Other than that, I'm ready to hit the road right now." She added, suddenly serious about getting on the road as she called it.

"Where's your rig parked?"

"Out by the front driveway, only place I could find to park it."

"We'll meet you there in about fifteen minutes, okay." K'Narf said, looking at Toby to see if he agreed.

"I get the sleeper." Toby said, stretching as he claimed it. Thinking as he said it of a long uninterrupted sleep, something he hadn't had since K'Narf had asked him to put a full purchasing department in place. A purchasing department that no longer needed his attention on a daily basis, he thought, feeling a well-deserved sense of satisfaction.

"Listening to Toby's slow steady breathing, K'narf turned to look at DeeDee's face all lit up by the lights from her big rig's dashboard and asked. "How are you and Karl getting along?"

"Is he asleep?" DeeDee asked, nodding at the sleeper behind her.

"Looks like it, he's all curled up with his back to us and he gives out with a little snort every now and then, why?" K'Narf asked, glancing back at Toby's sleeping form.

"Just didn't think he needed to hear what I had to say about Karl, that's all." DeeDee said, looking over at K'Narf.

"What's wrong with Karl?" K'Narf asked, remembering being worried about putting Karl and

DeeDee together, even if it was just to share information.

"Oh, nothing's wrong with Karl, just the opposite in fact. He's doing a fantastic job and everyone likes him including all the newbies." DeeDee said, her face lighting up as she mentioned how everyone liked Karl.

"Everyone likes him?" K'Narf asked, picking up a strange idea from the way DeeDee was talking about Karl. "How about you?" He asked, carefully watching DeeDee's face as he asked the question.

"Yeah, me too, but I don't think he likes me very much." DeeDee said, her face showing her disappointment even in the dim light of the big rig's cab.

"Why?"

"We didn't exactly hit it off at first." DeeDee said, looking over at K'Narf in a way that told him she had done something she shouldn't have.

"What did you do?"

"Got mad and flashed him an image of his body on a gurney getting shocked by stem paddles." DeeDee said, refusing to look at K'Narf.

"And what did he do?"

"Flashed me an image of my body as part of a pile of corpses. Turns out that the factory he worked at during the war was bombed and he was one of the ones assigned to haul out the bodies. Seems he's seen as much blood and gore as I have." DeeDee said, her voice getting very soft. "After that he started calling me 'shavetail', whatever that means."

"Are the two of you on speaking terms now?" K'Narf asked, beginning to see the problem.

"Yes, but no matter how nice I try to be, he still calls me 'shavetail'." DeeDee said, her voice quivering just enough for K'Narf to hear it. "Do you know what shavetail means, I don't, but everyone else seems to?" She asked, looking over at K'Narf.

"As a matter of fact, I do. My grandfather called me that for the first couple of years I worked on the farm with him. Looked it up in the library later on, wanted to know what it meant too." K'Narf said, looking over at DeeDee and feeling a little bit of her confusion and pain.

"And?" DeeDee asked, urging K'Narf to tell her what it meant.

"Well, in the beginning it meant young unbroken army mules, they used to shave their tails as a warning that it would kick if given the chance."

"You mean he's been calling me a mule!" DeeDee exclaimed, making her words a statement rather than a question.

"No, no, that's what it meant in the first world war, by the time the second world war rolled around it meant an untried second lieutenant, especially one recently out of school. A greenhorn, a tenderfoot, someone you need to look out for because they might do something, uh, unexpected." K'Narf stammered, having almost said "something stupid".

"No, he's been calling me a mule because he can't trust me not to kick him!" DeeDee said, sudden tears streaming down her face.

"You really like him don't you?" K'Narf asked, DeeDee's tears giving him a feeling of helplessness.

"Yeah, but what difference does it make, he ain't never going to like me now is he?" DeeDee said, wiping at her eyes as she looked over at K'Narf.

"I'm not real good at things like this, would you mind if I flashed T.J. and told her what's going on?" K'Narf asked, thinking T.J. would know what to do.

"I don't know, doesn't seem right bothering her with my problems." DeeDee said, wiping at her eyes again. "Guess it'd be all right though, if you think she wouldn't mind."

"Give me a minute." K'Narf said, closing his eyes as he dialed T.J.'s frequency.

"Well, what did she say?" DeeDee asked, seeing K'Narf open his eyes and knowing that he was no longer in touch with T.J.

"She wants you to flash her and Wanda on this frequency." K'Narf said, mentally sending DeeDee the frequency that T.J. had given him. "She seemed rather eager to help." He added, thinking that his sweet little wife had actually drooled when he told her what was going on.

"They really want to talk to me, I mean help me, really?" DeeDee asked, obviously pleased by what K'Narf had said.

"Yes, and while you three are scheming I'm going to join Toby and get some sleep." K'Narf said, consciously relaxing as much as four G's of constant thrust would allow. "Wake us before you hit mid-point. We'd rather be awake when you flip this thing over and reverse thrust." He added, as he closed his eyes.

"T.J." K'Narf flashed, hoping he wouldn't be waking her up. "Where are you?"

"I'm driving a lead Limo and we're on our way back to Sutton Wells why, what is it?"

"What did you decide about DeeDee?"

"We're getting together a plan, why?"

"Who's we?" K'Narf asked, sensing that T.J. didn't want to tell him.

"Are they awake?" T.J. asked, meaning DeeDee and Toby.

"Why, we're flashing."

"I know, I just don't want DeeDee to think we're talking about her."

"No, they're both out like a light, now what's going on and who's we?" K'Narf asked a second time.

"Did DeeDee tell you about that big commissary Karl built for her?" T.J. asked, ignoring K'Narf's question.

"Yes, can you believe Redwing running something like that?" K'Narf asked, realizing that T.J. wasn't going to tell him what they were up to. She doesn't trust me, K'Narf thought, surprised by the idea that he wasn't trusted.

"You gave her status when you told her to sit down and join the meeting, and now everyone who sees the emergency meeting minutes looks up to her." T.J. flashed, again totally ignoring K'Narf's question.

"Status?" K'Narf asked, not real sure what T.J. meant by status.

"My dear sweet K'Narf, God gave you an over abundance of status, but he also blinded you to the fact that you have it which is one of the reasons I love you so much." T.J. said, sending K'Narf a little red heart with "I love you" written on it.

"What do you mean, I don't understand what you're talking about." K'narf flashed, forgetting all about his original question.

"Sweetheart, remember when you *suggested* it might be a good idea to bring the newbies up to the Whale for orientation in a stretch Limo. I know you said it mostly to tease me, but there are now nine stretch Limo's going to the Whale loaded with newbies six days a week." T.J. flashed, stressing the word "*suggested*'" in her flash.

"Yes, so what's that got to do with this 'status' stuff?" K'Narf asked, getting more confused.

"Think my sweet husband! Status is given to people that make things happen, and you really, really do make things happen! Remember how you '*suggested*' it would be a good idea to buy an abandoned factory so we would have a place to build our fighters. Well, there are now slightly more than three thousand people down at the Sutton Wells factory working twenty four hours a day, six days a week doing just that." T.J. said, half bragging on him and half explaining to him the power of his words.

"Well, it was a good idea." K'Narf said, knowing how pleased he had been when it had all worked out.

"Yes, and it was your idea. Look at what happened when Brent threw you the proverbial hot potato in the middle of *his* emergency meeting." T.J. flashed, stressing the word "*his*" when she spoke of Brent. "You took it and told him to take that downed Snark ore ship, get it into space and convert it into a state-of-the-art command center. Brent and all of his pilots are up there right now along with over a hundred communication and wiring specialists doing just that." She added, trying to get K'Narf to understand the power of his words.

"How do you know that?"

"Little eagle and I keep in touch." T.J. flashed, letting K'Narf know that she had been keeping herself abreast of what was going on, unlike Toby and him. "Then you asked Mitch if he could run the factory if you stole Karl away. He said yes and now all those men working in the factory take his orders without question." She added, getting slightly frustrated with K'Narf's denseness.

"Yes, I know. I just spent the last six weeks helping him redesign the factory workflow. We cut the time down by six-point-two hours per completed fighter." K'Narf flashed, remembering the feeling of satisfaction he had felt when the final numbers came in. "That's a little more than a full shift." He added, feeling that sense of satisfaction all over again.

"Yes sweetheart, I know how you feel, but don't you see. You also asked Karl if he could put a dome on that Snark mine and now he's done it, along with the help of almost ten thousand men and women who have migrated up there to help him, all because you thought it would be a good idea, your good idea!" T.J. flashed, beginning to think it was a lost cause to make K'Narf conscious of the power of his words.

"You're saying all these things are happening just because I suggested them?" K'Narf asked, beginning to see what T.J. was driving at.

"Yes, and when you asked DeeDee to set up a commissary for coffee and sweet rolls, it blossomed into a building that can feed up to five hundred people per hour and house a staff of a hundred and three cooks and line servers with seven more being added this week." T.J. said, thinking she was finally getting through.

"How do you know that?" K'Narf asked, suddenly realizing that none of this stuff that had him and Toby on edge was news to T.J. "You knew about all those expenditures too, didn't you?" He asked, as the realization hit him that none of this was a surprise to T.J.

"You asked me to keep sufficient funds in the general account to cover whatever costs were incurred, didn't you?"

"Well, yes, I uh, guess I did at that." K'Narf stammered, suddenly seeing what T.J. had been trying to tell him about the power of ideas when they get translated into action, especially his ideas. "Now I'm nervous, maybe you shouldn't have told me about all this stuff." He added, feeling a knot developing in the pit of his stomach.

"No! You must not become nervous, just cautious! You are a leader and leaders by the very nature of their position carry a burden of responsibility!" T.J. flashed, feeling that finally K'Narf was beginning to understand what God had placed on his shoulders.

"Okay, but if I get even the least bit out of line, I'll depend on you to kick me in the shins!" K'Narf flashed, the knot in his stomach softening just a little as he quickly changed the subject. "How are you feeling, you're not getting nauseous anymore are you?"

"No, I'm no longer getting nauseous." T.J. flashed, unable to hide the worry that had been troubling her for the last week.

"What's wrong?" K'Narf asked. "Come on, tell me, I can feel your nervousness so don't try and hide anything from me." He added, knowing that T.J. knew he could feel her feelings under certain conditions, like right now.

"It's just that I should be showing a least a little bump by now." T.J. flashed, knowing that it could be just that she was a little over anxious to start showing.

"Is that all, I thought it was something serious." K'Narf flashed, trying in his own way to comfort T.J.

"You'll think not serious, if your first child turns out to be a worthless one." T.J. said, flashing back at K'Narf what was really worrying her.

"Worthless one, what do you mean?"

"Your first born should be a son, not a girl." T.J. explained, falling back on the culture she had been raised in.

"If it's my little girl she definitely won't be worthless!" K'narf flashed, defending his yet unborn baby girl. "Why do you think it might be a girl anyway?"

"I just think I should be further along than I am, that's all, just a little bigger by this time." T.J. flashed, letting K'Narf feel the worry that was in her heart as she changed the subject this time. "When do you expect to get there?"

"I can see Pluto coming into view now." K'Narf flashed, showing T.J. the planet Pluto coming up in the large windows of DeeDee's tractor-trailer. "Guess I ought to wake them up."

"Well, I need to be going too, I love you, and be careful!" T.J. said, as she flashed him another small pink heart with the words "I love you" printed on it and then cut the connection.

Chapter Twelve

"DeeDee, wake up! We're here!" K'Narf said, reaching over and shaking DeeDee's arm.

"Oh uh, okay!" DeeDee said, shaking herself as she straightened up in her big rig captain's chair. "Need to let them know we've arrived." She added, yawning as K'Narf reached back and shook Toby's leg waking him up too.

"They want us all to tune in to the common frequency." DeeDee said, moments later.

"Mister Etaguf, are you there?" A familiar voice asked.

"Yes, and so is Toby Zebulon, I hope you don't mind our dropping in like this, it's Karl isn't it?"

"Yes, and I've put out the word so everyone's tuning in. Okay everyone, you know what to do so let's get to it!" Karl ordered with such an air of authority that you could almost see everyone jumping up and getting to it, whatever "it" was.

"You want me to bring'em in?" DeeDee asked, as she carefully maneuvered her rig in to a high orbit.

"No! Mister Etaguf, you asked us to not only cover this old mine, but to camouflage her too. Well, we think we did a real bang up job of doing just that, so we'd like to see if you can find the mine and how to git into her too." Karl said, getting a half a dozen clapping hands flashed onto the common frequency letting K'Narf and Toby know that just about everyone was tuned in. "That'll give us a little time to get ready for your visit so to speak, and Shavetail, if you give us away I'll tie a not in your tail that'll take a week to git undone!" He added, speaking directly to DeeDee.

"Not if I kick you clear into next week first!" DeeDee said, flashing a caricature of a mule with a shaved tail kicking Karl clean off the planet.

"Chalk one up for DeeDee." An unidentified voice said, flashing a big neon checkmark.

"Maybe I ought to just take you over my knee and give you a few good swats instead." Karl growled.

"Not unless you change your name to McLintock and start wearing ten gallon hats you won't!" DeeDee fired back.

"Another one for DeeDee." The same unidentified voice said, this time flashing a double checkmark.

"Do I hear a volunteer for a double shift?" Karl asked, trying to identify the kibitzer.

"Now Karl darling, don't go picking on the fellows just because you're a little slow on the up take." DeeDee said, in her sweetest voice.

"Karl darling! Better make that three for DeeDee and I'm outta here, said the unidentified voice to a whole chorus of clapping hands.

"You just make sure Mister Etaguf finds us on his own, you hear me!"

"Yes sweetheart, whatever you say." DeeDee said, as she broke the connection.

You can almost see his face turning red, K'narf thought, looking over at the grinning DeeDee. "Did T.J. tell you to do that?"

"Tell me what?" DeeDee asked, trying to look innocent.

"What's she doing?" Toby asked, trying to figure out what was going on.

"T.J. and Wanda are giving her advice on how to get Karl."

"Sounds like something T.J. might come up with." Toby said, looking over at the smiling DeeDee.

"What do you mean?"

"Any woman down there." Toby said, pointing at Pluto. "Who heard DeeDee call Karl darling and sweetheart won't let him within talking distance now that DeeDee has put her brand on him." Toby said, noticing how DeeDee choked up when he mentioned branding Karl. "You might say that Karl has been taken off the market." He added, grinning at the still choking DeeDee, knowing that he had guessed right.

"Okay, enough foolishness, let's get down there and take a look at what they've done." K'Narf said, ignoring the still sputtering DeeDee as he turned his attention to Pluto.

"See those three peaks, drop down to that small crater at their base. Yes, that's the cave where we stored the library cylinders." K'Narf said, moments later as he pointed at the small cave in the side of the crater. "Now the mine is almost directly opposite this crater." He added, pointing at the way he wanted DeeDee to go.

"How many times have you been here?" Toby asked, as DeeDee faithfully followed K'Narf's directions.

"Twice, once when I accidentally found their mine, and the second time when T.J. and I attacked them." K'Narf said, watching as the terrain rolled away under DeeDee's big rig. "There, that ridge coming up, the mine's just on the other side of it." He added, as they cleared the ridge and flew over it.

"You sure you got the right ridge?" Toby asked, looking at the ground passing by underneath them. "There's never been a mine around here, you can tell by the age of the rocks. That's something that can't be faked." He added, noticing how the terrain had a time worn look to it that newly turned stone or rock wouldn't have.

"No, it's here, right underneath our feet so to speak. I've just got to find the way in." K'Narf said, looking carefully at the area where the mine used to be, noticing as he did that there was a noticeable bulge in the center.

"You're sure about this?" Toby asked, still looking for some sign of newly turned rock or soil.

"Go over toward that ravine off to the right." K'Narf said, pointing at a small dark cut in the side of what he thought should be the dome.

"Yes sir." DeeDee said, moving her rig toward the ravine K'Narf had pointed out.

"Okay, take her down into it and turn your headlights on." K'Narf said, looking at a dark shadowed area in the rock face ahead of them. "Okay, move forward very slowly."

"Yes sir."

"If you're gon'na run us into a mountain, do you mind if I put my helmet on?" Toby asked, reaching back into the sleeper for his fishbowl.

"Will light penetrate this force field you told me about?" K'Narf asked, looking at the blackness in front of them.

"No sir, I don't believe it can." DeeDee said, stopping her rig inches away from what appeared to be a black curtain that seemed to be absorbing all the light from their headlights.

"Then take us through DeeDee." K'Narf said, looking over at DeeDee.

"Yes sir!" DeeDee said, grinning from ear to ear as she eased her rig into the blackness of the curtain in front of them, instantly bringing them into an enormous area with giant steel columns marching in stately circles ending in a huge cleared area down in the center that was lit up like a Las Vegas Casino.

As they descended toward the lights it became obvious that there were thousands of people all waving and yelling at them to come down, and right in front of where they wanted them to land was a huge banner spelling out the words 'WELCOME TO ETAGUF CITY'.

"Etaguf City!" K'Narf exclaimed. "I'll kill him, and there ain't no T.J. here to stop me this time." He added, as DeeDee set her rig down in what appeared to be a mob of applauding people; people all wearing silver collars and blue green space suits along with every description and color of coats, shirts, skirts, and hats you could imagine.

"If they don't need helmets, why are they still wearing their space suits?" Toby asked, looking at the wildly applauding people that were now mobbing DeeDee's rig.

"Haven't got the sewer system in operation yet." DeeDee said, as she opened the door to her rig, and allowed the deafening sound of applauding hands into the cab.

"That was awful fast, did you show him where it was?" Karl asked, as he leaped up on the running board and stuck his head in.

"No, he figured it all out by himself, didn't pay any attention to all that acid washing you did to those rocks either." DeeDee said, firing back at Karl.

"Acid washing, so that's how you did it." Toby said, jumping into the conversation as he figured out how they had artificially aged the rocks to look like they'd been there forever.

"Mister Etaguf, Mister Zebulon, welcome to Etaguf City!" Karl said, finally turning his attention to K'Narf and Toby.

"Thanks Karl, but we sweated too many hours together for you to be calling me Mister anything, name's K'Narf remember." K'Narf said, climbing out, followed immediately by Toby.

"Yeah, sorry about that, everyone here calls you that and I guess I picked up the habit too." Karl explained, shaking K'Narf's hand and then Toby's.

"Git clear of my rig, got'ta run these supplies up to the commissary." DeeDee said, motioning with her left hand for everyone to get clear as she lifted her rig and headed for a large building way up on the fifth level.

"Don't have any fancy speeches lined up or anything like that, everyone just thought it wouldn't be proper not to welcome the man they named the city after." Karl said, as everyone began to crowd up around them now that DeeDee's rig was out of the way. "You probably ought to go shake a few hands, they been working their tails off building this here place for you." He added, as he pushed K'Narf toward the people crowding in around them.

"They named the city, not Brent?" Toby asked, watching as K'Narf was swallowed up by a mob of people all trying to shake his hand at the same time.

"Yeah, name just kind of evolved, he was the one asked me to build the place and everyone knew it, so in their minds they were working for him as well as me. Later, when we began to realize that it was going to be more than just a second factory site they started calling it Etaguf City, why?"

"Oh nothing, just figured Brent might have had something to do with it." Toby said, thinking K'Narf was going to have a sore hand as he watched a rather big man shaking the fool out of it. "By the way, got a little heads up for you."

"What's that?" Karl asked, turning to look at Toby.

"Seems DeeDee has set her cap for you."

"How do you figure that?"

"Picked it up on the way out here, and she's got some of her girlfriends helping her if I'm understanding it right."

"So that's why she called me 'darling' and 'sweetheart'." Karl said, frowning for a second then getting a big grin on his face. "Well, we'll just have to see about that."

"Think we ought to go rescue him before they shake his arm off?" Toby asked, pointing at K'Narf who was having his hand shaken by another rather large gentleman.

"Yeah, let's go git' em." Karl agreed, pushing his way into the crowd surrounding K'Narf. "All right, that's enough, let the man breath, go on now, git back to work or to bed or whatever else you're supposed to be doing. I mean it now, go on git!" Karl said, raising his voice as he physically started shooing people away from the frazzled looking K'Narf.

"You all right?" Toby asked, seeing the dazed look on K'Narf's face.

"I don't know." K'Narf said, rubbing his right arm.

"Just part of a hero's job being adored." Karl said, grinning at K'Narf. "Come on, let's git up to Redwing's commissary and get us a cup of coffee. It'll give me a chance to explain the layout and everything on the way.

"Sounds good to me." Toby said, stepping over to the other side of K'Narf. Effectively sandwiching K'Narf in between himself and Karl as they started walking.

"First off as you can see there's seventy two columns, we put flood lights on the bases to help light the place up. It's still dark between the columns, but at least you can get a feel for the size of the place." Karl said, waving his arm in a full circle. "Center circle starts with eight columns as you can see, but by the time we got out to the third, fourth and fifth levels we had to add eight more to each circle. Each column segment has a diameter of fifteen feet by twenty feet high with three-inch thick steel walls. Base segments are pre-designed for sewer lines, water and power, with power being prewired into all riser segments.

If you look up you can see that the tops of all the columns are connected with circular steel bands of three-inch thick steel, and cross connected with the same material. Center of the dome was designed like a typical arch just a whole lot bigger. The whole structure was covered over with half-inch thick steel sheeting which was then roofed over with three-inch thick steel slabs just like you would shingle a house.

"Why did you need the shingles, isn't that a lot of extra weight?" K'Narf asked, thinking of the tremendous weight over their heads.

"What do you normally weigh?" Karl asked, turning to look at K'Narf.

"Around two-ten." K'Narf answered, meaning two hundred and ten pounds.

"Two-ten times point one nine eight five would make your weight up here around forty one and a half pounds. That brings your weight down, but it does not diminish your physical strength."

"Strength, why'd you need more strength?" Toby asked, thinking that half-inch steel sheeting would have been more than enough.

"Meteors, there's no atmosphere out here so they pretty much come straight in." Karl said, looking around K'Narf at Toby.

"How much danger is there?" K'Narf asked, realizing what a meteor hit could do.

"Well, we checked the planet over and could find no indication of any recent meteor strikes, but there was a lot of evidence of micro-meteor hits. The deepest penetration we found was about six to eight feet, we decided to increase the strength of the whole dome by three inches of steel and then we covered it with approximately twenty-five to thirty-five feet of dirt and rock. Then we acid washed everything to age it and you saw the results when you flew in today." Karl said, obviously proud of what had been done.

"I'll admit it had me fooled." Toby said, thinking that three inches of extra steel over his head wasn't such a bad thing.

"Would've fooled me too, if I hadn't known exactly where it was and the fact that you dared me to find it." K'Narf said, grinning at Karl. "Though I doubt I'd have found the entrance if DeeDee hadn't told me about the new force field you were using as an airlock."

"I knew it! Bless that girl and her mouth!" Karl exclaimed, thinking DeeDee had told K'Narf where the entrance was.

"No, no, she mentioned that on the way out, I just put two and two together and figured it out. She didn't tell me anything after you asked her not to." K'Narf explained, trying to get DeeDee out of trouble.

"Well anyway, we got the new sewer system all hooked up and in place and we'll have our new state-of-the-art waste treatment plant operational by next week, then everyone can get out of these things." Karl said, tugging at his space suit.

"You have a treatment plant?" Toby asked, looking around for a sign of such a plant.

"Sorry, yes. You're standing over the central sewer hub right now. It connects to all eight of the mains coming in from the columns." Karl said, drawing a line with his finger from the nearest column to the ground they were standing on. "It's buried about fifteen feet below your feet and it leads off that way." He added, as he pointed off to his left.

"I didn't see anything out that way when we flew in." K'Narf said, trying to remember anything that might have been a treatment plant.

"No, you wouldn't have, that's where we got most of the dirt and rock for the roof. We dug a huge pit a little over two hundred feet across and about ninety feet deep and roofed it over just like the dome here. Tunnel over there leads you to it." He said, pointing at a barely visible opening in the first level's stack block wall.

"And that." K'Narf asked, pointing at what appeared to be the making of a small lake bed in the center of the bottom area."

"That's going to be the center piece for this whole green zone according to Rudy, gon'na have a fountain sending up a stream of water about a hundred feet high. He said he wanted it to look like some place in Arizona called Fountain Hills whatever that is." Karl said, looking at what K'Narf had pointed out and asked about.

"Rudy, what's he got to do with any of this?" K'Narf asked, surprised by the mention of Rudy's name.

"Rudy's in charge of planning out and landscaping all of the green zones. I thought you knew that." Karl said, surprised by K'Narf's question.

"No, I didn't know that, and what's a green zone?"

"Well, this whole area your standing in for starters, Rudy says it's going to be a park with lots of trees for everyone to enjoy, as well as a year around garden for veggies and stuff like that." Karl said, then pointing at the stairs they were coming up to he continued. "See how we've built these stairs into the stack wall behind each of the columns. They're fifty feet wide and represent a fifty-foot wide strip of ground that leads to the next column and the stairs behind it. All the columns are set up the same way. The sewer, water and electrical lines are buried in these strips between the columns, that's why they are going to be walkways, if we have to do any repairs in the future it will be easy to dig them up and Rudy has claimed a thirty five foot wide strip on each side of all of them for future green zones. He says if he can get enough trees growing we may not need to depend on outside sources for our oxygen."

"Rudy said that?"

"Yes, said he was your Head Gardener and kinda took over and I wasn't about to argue with him, besides, everything he said made sense." Karl said, wondering if he should have taken Rudy at his word. "He is your Head Gardener, isn't he?"

"Yes, as a matter of fact he is, I just haven't had a chance to talk to him lately. Been working with Mitch Murphy for the last six weeks and I'm a little out of touch with what's been going on, that's all." K'Narf said, thinking of what T.J. had said about the power of his words and how he had agreed with Rudy that his title was that of Head Gardener, but not in his wildest imaginings had he ever thought it would lead to something like this.

"Those stack walls seem pretty formidable." Toby said, changing the subject as they approached the huge stones that made up the fifty-foot high wall.

"Yes, they were one of the first things we did after we got up here. Dug a trench at the base of all the circles and filled them with pea gravel, then stacked the walls up before anything else could happen."

"Why, what happened?" K'Narf asked, seeing the look on Karl's face change.

"Sorry, keep forgetting that you're new up here. Dirt edge broke off with one of my men standing on it." Karl said, pointing up at the top of the wall. "The fall broke his helmet when he hit bottom. He was dead by the time we got to him. Figured there might be more as time goes on so we dedicated a little cemetery and buried him in it about a mile from here.

"I'm sorry, I didn't know." K'Narf said, feeling like this is something he should have known about.

"Speaking of life and death, rumor has it that you're going to be a papa pretty soon, any truth to it?" Karl asked, turning to look at K'Narf as they started up the first flight of stairs.

"Tell him how she told you." Toby said. "T.J. did it up right." He added, looking around K'Narf at Karl with a big grin on his face.

"Yeah, yeah, and don't leave out any of the juicy details." Karl said, just as they got to the top of the stairs and looked at the hundred yards of raw dirt separating them from the next pillar in line. A shiny steel monolith lit up by ground level floodlights with it's top disappearing into the darkness of the massive domed roof.

"Not that much to tell." K'Narf said, digging an elbow into Toby's side in an effort to shut him up. "She made me get dressed up in a real fancy tuxedo, you know, one of those ones with a ruffly shirt and a cumber bun, then she put on one of the most beautiful white satin dresses I've ever seen." K'Narf said, pausing as he mentally remembered how beautiful T.J. had looked.

"Tell him about her jewelry." Toby said, totally ignoring K'Narf's jab in the ribs.

"She had gone to what she called her vault and brought out an unbelievable set of antique pearls. I'd be willing to bet those pearls haven't seen the light of day for more than a century. She worked them into her hair, put them around her neck and hung them from her ears. The effect with that white dress was indescribable." K'narf said, as he flashed them a picture of what T.J. had looked like.

"Lord have mercy!" Karl exclaimed, stopping as he took a full minute to examine the picture K'Narf had sent him. "And then?"

"She took me to what must have been the Taj Mahal of oriental restaurants where they bowed and scrapped to her like she was the Empress herself. The service and food were one star above five and then she told me." K'Narf said, thinking he said enough.

"Tell him how she told you." Toby said, egging K'Narf on.

"What does he mean?" Karl asked, fully into K'Narf's story.

"Okay, okay, she had brought a small white jeweled box with her and after we finished eating she opened it and took out a very delicate looking paper doll that had been folded out of a real fine paper you could almost see through." K'Narf said, continuing at Toby's insistence.

"It was rice paper." Toby said, helping K'Narf out with his description.

"Yes, yes, rice paper." K'Narf said, elbowing Toby again. "Then she gave me a kind of curtsy as she handed me the doll and said 'This represents the child that my body will soon present to you, may it bless you with eternal joy and happiness. It took me a full minute to figure out that she was telling me she was pregnant. Her face had a smile on it that lit up the whole restaurant as well as my heart, and we spent the rest of the evening holding hands and that's all I'm going to tell you." K'Narf said, giving Toby another dig in the ribs for good measure as they reached the second level stairs.

"Sounds like something out of a story book." Karl said, thinking that being bonded with a pretty woman might not be such a bad thing.

Lost in private thoughts, they walked in silence until they reached the fifth level then K'Narf saw all the tents. "What are those?" He asked, as he looked around and began to see tents looming out of the dark areas between the brightly lit pillars in every direction.

"Well, except for the ladies living over the commissary, everyone else lives in tents. It's the best we can do until we get some apartment buildings built and that probably won't happen until this little fight we're about to get into is finished." Karl said, heading for the building that was obviously the commissary. They were close enough now for K'Narf to read the sign over the double glass doors that read "REDWING'S COMMISSARY" with the words "Open 24 Hours A Day" spelled out in smaller letters underneath it.

Chapter Thirteen

"What in the world!" K'Narf exclaimed as Karl opened the door for him and he stepped into a very busy and noisy Commissary.

"Something isn't it." Karl said, stepping in behind Toby. "We couldn't do much about the layout since we didn't have room for booths and fancy stuff like that. Had to cram in as many tables as possible and the conveyor belt that takes the dirty dishes back to the kitchen couldn't go anywhere else but up against that far wall, but every thing else you see except the kitchen proper we owe to T.J. and Irena."

"T.J. and Irena!" K'Narf exclaimed, beating Toby by a millisecond, just as everyone started standing up and applauding as they began to recognize him. "How long am I going to have to put up with this?" He asked, turning to look at Karl.

"Probably as long as you have that face." Karl said, grinning.

"It won't work." Toby said, shaking his head.

"What won't work?" Karl asked, wondering what he had missed.

"He's thinking of changing his face." Toby said, smiling at Karl. "But the first time someone spotted him with T.J. they'd know who he was again so it wouldn't work." He added, looking seriously at K'Narf as one by one the hundreds of diners sat back down and returned to the business of eating.

"More coming in, we'd better get in line." Karl said, leading them over to the right side of the room where a fancy chrome stand with a velvet rope divider separated the diners from those waiting in line.

"T.J. and Irena set all this up?" K'Narf asked, stunned by the luxury of the dining area and by the fact that T.J. had been somehow involved.

"Are those real linen tablecloths?" Toby asked, looking a little closer at the nearest table.

"Everything here is real, well almost everything." Karl said, rethinking what he had just said.

What do you mean almost?"

"Well, Redwing didn't want sterling because of all the trouble it'd be to keep it polished so they bought stainless steel flatware in a pattern called Michelangelo instead, and the flowers are some kind of fancy silk they found in China, but everything else is real. By everything else I mean all of the glasses, creamers, water pitchers and all that other stuff are real European crystal, and those gold leaf dishes come from a company in China. I think T.J. owns it, or at least part of it and it's supposed to be the same company that used to make the china for the Emperor in the old days." Karl said, bragging about how T.J. and Irena had put together the table settings.

"And the chandeliers?" K'Narf asked, looking at how a beautiful two tiered crystal chandelier centered above each table made the facets of a matching three tiered crystal Lazy Susan sparkle and glisten as it was turned by diners looking for various condiments; each one topped by a different arrangement of silk flowers and greenery.

"Yep, all of it came from hi-falootin' companies in Europe." Karl bragged, as he urged them forward in order to keep up with the moving line, waving back at some of the diners as they waved at him.

"How do you know that?" Toby asked, still a little skeptical about what he was seeing.

"Seen the labels on the crates when they came in, that's how!" Karl said, raising his voice at his word being questioned.

"And the wood floor, is it real too?" K'Narf asked, looking down at the rich texture of the wood under his feet.

"Naw, that's a new artificial material that you couldn't scratch if you tried. Figured we'd need the toughest material we could find considering the beating a floor in this place was going to take, but them tables are real. Each pedestal table and its eight heavy-duty chairs are made out of hand carved teak wood. I wanted square ones, but the ladies insisted on round ones. We could only fit thirty eight of them in because they needed a little space over by where you have to put your tray on the conveyer belt, otherwise we could have got forty in.

"What's that thing you have screwed to the back of the chairs?" K'Narf asked, looking at an empty chair just on the other side of the rope."

"Those, uh, well, the table was way too pretty to put the tray on so we jury rigged up a tray holder and screwed it to the back of each of the chairs. Ladies got a little upset, but we didn't know what else to do. They finally got over their mad and use'em too, though." Karl said, looking a little sheepish.

"I'll bet that messed up T.J.'s Yin and Yang!" Toby said, looking over at Karl.

"What did you mean when you said 'except the kitchen'?" K'Narf asked, looking back into what appeared to be an unusually large and spotless kitchen.

"Redwing and her cooks designed that; T.J. gave her Carte Blanche so they went out and bought the best there was in everything from knives to garbage cans, everything's stainless steel and they keep it polished to a tee." Karl said, glancing at the spotless kitchen, at least the parts they could see from the serving line.

"Very nice." K'Narf said, wondering what had come over T.J. She was generous to a fault, he knew, but this was just a little over the top even for her.

"Hey, what do you think of the mural Debbie's painting for us?" Karl asked, changing the subject as he turned and faced the wall they were standing next too.

"Debbie's been up here too?"

"Debbie, yeah, she helped set everything up in here. After they were done a bunch of them were all setting around talking about what they thought the city should look like and she started drawing sketches; week later she hitched a ride up with one of my truckers along with all of her paints and brushes and started doing this. She left to get more paint, but we expect her back in a couple of days." Karl said, pointing at the half finished mural on the wall behind them.

"Thought that bottom area was supposed to

be a green zone." K'Narf asked, looking at an area that represented the center all sketched in with shops and what looked like a big bowling alley and theater complex.

"It is, your looking at the second deck. See that water fountain she's drawn in squirting up in the center, that's coming from that big fountain in the green zone underneath. Same thing you're seeing over here." Karl said, moving a little farther up the line and pointing at a section of the mural that had been finished. "This shopping mall is built over the combined walkway and green zones underneath it. When you combine the walkway and green zone spaces you have an area one hundred and twenty feet wide by seven hundred feet long, more than enough room for little strip malls like the one she painted here." Karl said, pointing out how Debbie had painted in little shops lining both sides of a second level walkway.

"Then those shops are literally built over the green zones aren't they?" Toby asked, looking at how the opposing rows of shops matched the area of the green zones below them.

"Yes, we were going to have to build an overhead support to mount the thousands of grow lights on anyway, so why not make it strong enough to support a second level above it?" Karl asked, explaining the logic behind the design.

"And this building?" K'Narf asked, looking at the massive building with it's many windows and mini balconies painted in next to the row of shops.

"An apartment building. Most of the city will be made up of buildings like that with the tallest ones being down on level one and the shortest ones up on levels five and six." Karl said, moving forward to keep up with the fast moving line.

"And these?" K'Narf asked, pointing at what appeared to be suspended walkways.

"Exactly what they look like. All of the apartment buildings will be interconnected with multilevel walkways." Karl said, shooing them forward again as they reached the area where the trays were stacked.

"Looks kind of flimsy to me." Toby said, taking a last look at the feathery looking walkways.

"Newer and stronger materials with less weight to support." Karl said, as K'Narf and Toby picked up their trays. "That walkway would hold an elephant up here." He added, as he picked up his own tray.

"How many people do you think this city could hold if it was finished out the way Debbie is painting it?" K'Narf asked, picking up a napkin wrapped set of Michelangelo flatware.

"Best we could figure without actually drawing everything out was just a shade over a million, why?" Karl asked, picking up his own flatware.

"A million!" Toby exclaimed. "You've got to be kidding."

"Nope, if not more. Uh oh, look who's here." Karl said, looking at the entrance door.

"Who?" Toby asked, turning around.

"Brent and his scallywags." Karl said, pointing at Brent and four men who had just came in with him. "Redwing will be thrilled." He added, grinning.

"Why did you call them scallywags and why will Redwing be thrilled?" K'Narf asked, picking up on the word scallywag.

"You have to understand that we don't have any form of entertainment up here but ourselves, so when Brent comes over from that ship of his every week and loads up on supplies from Redwing's pantry, it turns the place into a three ring circus." Karl said, smiling from ear to ear. "Go ahead and take whatever you want, I've already eaten so all I'm going to get is a piece of Maude's apple pie and a cup of coffee."

"You sure there's enough to choose from?" Toby asked, looking down the line at a hundred feet of self-serve dishes, some piping hot and some on beds of ice.

"It depends on whether you're coming off a shift or getting ready to go on one." Karl said, reaching in and taking a large slice of rich looking apple pie.

"What do you mean?

"Might be breakfast for you, or lunch or maybe dinner. So Redwing and her people make sure you have enough choices to accommodate whatever meal you may be eating, that's all." Karl said, noticing that K'Narf was starting to put stuff on his tray, but that Toby hadn't chosen anything yet.

"What do you mean a three ring circus?" K'Narf asked, ladling gravy onto his mashed potatoes.

"Brent and his pilots have their own galley out on that ship of theirs so they don't need to come over here to eat, but they still need supplies. DeeDee offered to make a monthly supply run over to them but Brent said no, he'd rather come over here on a weekly basis and take what they need from Redwing's pantry. That prompted Redwing to call him a pirate and the four cooks he brought with him scallywags. Anyway, the next week he showed up dressed as Captain Jack Sparrow, looked pretty much like him too." Karl said, nodding at Brent who was now standing at the end of the line.

"Captain Jack Sparrow, you don't mean Johnny Depp?"

"The one and the same."

"Who's Johnny Depp?" K'Narf asked, turning around to look at Toby and Karl.

"It's another of those movie things K'Narf, just forget it." Toby said, finally picking up a small bowl of steaming spinach. "And then what happened?"

"Well, what do pirates do?"

"They steal things?"

"Right, and that's exactly what he did. When we get up to the soup you'll see a big fancy silver dipper. He took it out of the soup, wiped it off and stuck in his belt like a sword and walked out with it along with some of her flatware and one of her flower bouquets."

"What happened then?" K'Narf asked, getting interested in Karl's story.

"She loaded up one of the vans with some of her prettiest ladies and paid him a visit. There's a lot of bonded women over there and they didn't like Redwing and her girls showing up, so they made sure she got all her stuff back and laid the law down with Brent. Redwing still doesn't trust him though; she'll be out here in a few minutes and you'll see what I mean."

"Which is exactly what Brent wants, isn't it?" K'narf asked, looking back at Karl.

"That's about the way we got it figured." Karl said, nodding agreement with K'Narf. "And here she comes!" He exclaimed, as they all turned to see Redwing and four ladies coming out of the back of the kitchen. "Got everything you want, then let's get some coffee and find a table, shows about to start." He added, setting his tray down in front of a row of coffee urns.

"Gentlemen, may we join you?" Brent asked, a few minutes later as he walked up to their table and asked permission to set down followed by all four of his cooks.

"It would be our pleasure." Karl said, as K'Narf and Toby nodded their heads in agreement, both of them having been caught with their mouths full.

"Thank you." Brent said, quickly unloading his tray and stashing it behind his chair. "I heard you were here and headed right over." He added, as he sat down and bowed his head for a moment of silent prayer before raising it and taking a sip of coffee.

"These gentlemen are my head cooks, Jim, Jack, Lenny and Benny, they came with me to pick up our weekly ration of supplies." Brent said, as each man in turn nodded at the mention of his name.

"Okay ladies you know your positions!" Redwing exclaimed, as she walked up to the table and stood directly behind Brent with each of her four ladies taking the same position behind his cooks. "Nothing leaves this building that they didn't bring in with them." She added, glaring down at the top of Brent's head.

"Mister Etaguf, you remember reading about the Kings and Lords of old who always had a personal attendant standing behind their chair during dinner just in case they might need something, well this is the same thing with a modern twist." Brent said, looking over at K'Narf.

"Oh, K'Narf, I uh, I mean Mister Etaguf. I'm so sorry, I didn't see you there; I, uh, didn't mean to interrupt your dinner. It's just that, oh, just don't let him steal anything!" Redwing exclaimed, embarrassed as she turned on her heal and headed back toward her kitchen followed by four very confused ladies.

"Now look what you've done, you've scared her off." Brent said, turning in his chair so he could watch the fast retreating Redwing.

"So, you two still haven't resolved your problems yet?" K'Narf asked, looking over at Brent.

"Oh, I'd say they're resolved, I just can't do anything about them yet." Brent said, smiling as he turned to face K'Narf. "By the way, I understand you did a real nice job in cutting six point two hours off the assembly time down at Sutton Wells." He added, letting both K'Narf and Toby know that he did not need an update on their activities.

"Thank you, after we take a look at the line Karl is setting up, I'd like to take a look at whatever it is that you've been doing. Would that be possible?" K'Narf asked, biting the last bit of meat off a fried lamb chop.

"Not only possible but you would be our honored guests. That's exactly why I'm here a day early, I wanted to make the invite personally." Brent said, giving K'Narf a slight bow as he turned and looked at Toby to make sure he knew the invitation included him as well. Then he turned to his food with a single mindedness that astonished even Toby.

"Thank you, your gesture is very much appreciated." K'Narf said, picking up his coffee cup as he leaned back in his chair.

"Karl, would you like to come along too, we'll be doing a little show and tell and you'd be more than welcome?" Brent asked, moments later as he leaned back in his chair, indicating that he too was done.

"Actually, I would very much like to come if you're sure it would be okay." Karl said, obviously pleased that he had been included.

"Good, give me a heads up about ten minutes before you're ready to come over and I'll send some of my best pilots over to pick you up."

"Some?" Toby asked, picking up on the plurality of the word "some".

"Don't worry!" Brent said, laughing at Toby. "This won't be a test ride, just a little hop out to our new command ship." He added, grinning at K'Narf and Toby as he stood up, letting them know that he also knew about their little test ride.

"Good." K'Narf said, knowing they didn't have anymore of DeeDee's pain pills with them.

"Oh darn, look what you did." Brent said, standing up. "You scared off our personal attendants and now we'll have to carry our own trays; probably have to load up our own supplies too." He added, smiling softly as he pulled his tray out from behind his chair and stacked it up with his dirty dishes. Then with a slight bow he headed for the conveyer belt followed by all four of his cooks.

"His cooks sure aren't a talkative bunch." Toby said, watching Brent thread his way through the tables between him and the far wall.

"Normally they are, but I think you two intimidate them. Anyway, we don't need them as long as we have Brent and Redwing to entertain us."

Karl said, as Toby turned around and looked at him.

"Yes, but then of course there's you and DeeDee standing in the wings so to speak." Toby said, referring to Karl and DeeDee's little tiff when they had landed.

"You're right about that." Karl said, shaking his head in feigned misery. "And I'd be willing to bet that someone has already snuck into my tent and measured one of my caps."

"Why would they do that?" K'Narf asked, not understanding what was going on.

"So they know what size ten gallon hat to order." Toby said, knowing instantly what Karl was talking about.

"Bet it's already on one of my rigs heading this way." Karl added, again shaking his head in feigned misery.

"Oh, I see what you mean." K'Narf said, not really seeing what they were talking about, but acting like he did. "Would you two give me a moment, I'd like to contact T.J. before we go take a look at your new line?"

"No, no, take all the time you need." Karl said, leaning back in his chair.

"T.J., where are you?" K'Narf asked, hoping he wasn't waking her up as he flashed his question on T.J.'s personal frequency.

"At Sutton Wells, just got back, why, what's wrong?" T.J. flashed back.

"Nothing's wrong, just had dinner in the new commissary. Brent and four of his cooks joined us. They just left and we're getting ready to go look at Karl's new assembly line." K'Narf flashed, wondering if this was the right time to ask the question he wanted to ask.

"What is it K'Narf, what's bothering you?" T.J. asked, reading K'Narf's feelings much easier than he could read hers.

"Well, this place is beautiful, but you and Irena seemed to have went just a little bit overboard with it and I was wondering why, that's all." K'Narf flashed, asking his question anyway.

"Little Willie, that's why." T.J. flashed.

"I don't understand?"

"William 'Little Willie' Getz was one of the very first ones that I found and saved from a gambling syndicate fate. He was teaching High School English when the Lord touched him and pulled his filters, and he was one of the sweetest Christians you'd ever hope to meet, always willing to pull an extra shift for me if I asked."

"I still don't understand?" K'Narf flashed.

"Little Willie was the man who fell and broke his helmet." T.J. flashed, letting K'Narf feel her sadness at the loss of her friend.

"I'm so sorry, why didn't you tell me?"

"You were deeply involved in fixing that assembly line, and besides my problems could wait. As to why Irena and I did what we did, Little Willie's last meal was a cold sandwich sitting in the cab of one of Karl's rigs."

"Why?" K'Narf asked. "I mean why in one of Karl's rigs?"

"That's the only place he could go. He had to get into one of the rigs and pressurize the cabin before he could take his helmet off in order to eat or drink, and I felt that this was unacceptable for someone who was risking his life for us, and I had all this money and so I asked Irena to help me and we went shopping. More men or women may die up there but at least they'll know that they were appreciated!" T.J. said, beginning to tear up as she tried to explain what she was feeling.

"Okay, okay, I was just wondering." K'Narf flashed, not knowing what to say. He'd never felt T.J. this emotional before and wasn't sure what to do. "You were right about Brent, he isn't the same young man we forced to become our Head of Security." He added, trying to change the subject.

"Yes, he's much more the man now, and much less the teenager." T.J. flashed, agreeing with K'Narf as she got control of her emotions. "You say you're getting ready to go look at Karl's new assembly line?"

"Yes, and then we're all going to go out to see what Brent's done to that old Snark ship." K'Narf flashed, almost wishing he hadn't bothered T.J.

"Well, you go on then, and please flash me on your way back, okay?" T.J. asked, as K'Narf flashed her a small red heart with "I Love You" written on it as she broke the connection.

"Ready to go?" Karl asked, stretching as he scooted his chair back the minute K'Narf looked over at him letting him know that he was no longer in touch with T.J.

"Let's do it." K'Narf said, getting up and reaching behind his chair for his tray.

Chapter Fourteen

"Well, it's well lit, I'll give you that." K'Narf said, as they topped the stairs leading to level six, the topmost level, and he looked over at the area where the back wall and the roof touched.

"You can't do good work if you can't see what you're doing." Karl said, as they walked toward a brightly lit area snuggled up against the back wall, an area that looked to be at least twice the length of a football field.

"Is that a railroad track?" Toby asked, looking at what appeared to be a single set of train tracks running the entire length of the assembly area.

"In a sense yes." Karl said, pointing at the tracks. "Let me explain, it's really very simple. That track is what the assembly cradle rides on and there are nineteen assembly stations laid out along it. Once the work scheduled for a station is complete the Lead Inspector pushes a button and a green light comes on. This tells the line manager on duty that that station is ready to move. When all of the stations have a green light, then the order is given to move all of the assembly cradles ahead one station.

"How much work have you scheduled for

each station?"

"It varies, but we've tried to level load each station with the same amount of hours wherever possible." Karl said, knowing exactly what K'Narf was driving at.

"And the sub-assemblies?" K'Narf asked, knowing that had caused him and Mitch most of their headaches.

"Each assembly station has its own benches that supply the needed sub-assemblies for that particular station. If we're held up it'll be because of a lack of parts, not sub-assemblies." Karl said, looking directly at Toby.

"You'll have your parts." Toby said, squirming a little.

"When?"

"Kits are being put together now." Toby said, getting defensive. "You should have the first of them by this time next week." He added, praying he was right.

"What kind of a production rate will you get?" K'Narf asked, having been heavily involved with that very question for the last six weeks.

"We intend to fly a fully tested fighter off the end of the line every forty-eight hours, which in the time we have will allow us to produce one hundred and fifty fighters."

"One hundred and fifty! That's exactly what we figured we could produce at Sutton Wells." K'Narf said, surprised at the number since Sutton Wells had a three month head start on Karl and his team.

"Precisely, which put together with our production will give Brent a full compliment of three hundred fighters to go with those three hundred teams you told him to train, remember." Karl said, referring back to the meeting where K'Narf had told Brent to cut back his selection process to just three hundred teams. "You called them '*lone wolf*' teams remember, well, he's going to have a fighter to go with every one of them or my name isn't Karl Grauer!" Karl exclaimed, stressing the word "*lone wolf*" as he swung his arm to include the entire panorama of the brightly lit assembly area.

"Three hundred, that's still not very many when you think of what they'll have to go up against." Toby said, looking grim as he said it.

"It was enough for Gideon, it will be enough for us." K'narf said softly, pleased with what Karl had told him. "Let's give Brent a call, I think it's time to go see what miracles he has wrought with that old ore ship." He added, smiling at the look on Toby and Karl's face.

"You got it." Karl said, closing his eyes for just a few seconds. "Brent says they're already on their way. In fact, that should be them." He added, turning around as three Teardrop fighters floated up over the edge of level six and came slowly toward them.

"Mister Etaguf, Mister Zebulon, long time no see." Peter said, climbing down out of the first fighter to land. "Ready for another little ride?"

"Karl!" K'Narf and Toby exclaimed at the same time. "Peter here wants to give you a ride."

"Aw, come on. It wasn't that bad." Peter said, smiling from ear to ear.

"Not that bad, I couldn't walk right for three days." Toby said, involuntarily backing away from Peter as he approached with his hand out.

"Make that closer to four days." K'Narf said,

taking Peter's offered hand and shaking it.

"I take it you two know Pete." Karl said, noticing the reaction Peter had had on K'Narf and Toby.

"We've met before." Toby said, cautiously taking Peter's hand and shaking it.

"All kidding aside, this is just a little hop over to Big Mama and I promise you there won't be any fancy maneuvering." Peter said, getting serious.

"Big Mama?"

"That's what they call their command ship." Karl said, having heard it called that before as he stepped over to Peter's fighter.

"Ready to go?" Peter asked, seeing that the other two Teardrops had landed and their pilots had already thrown out the flexible ladders needed to climb up into the cockpits.

"How come we're getting into the front seat?" K'Narf asked, noticing that it was the front cockpit that had been opened for them.

"See better from the front seat and Brent wants to find out how long it takes you to spot Big Mama on our way in." Peter said, helping Karl as he started to climb the ladder he'd thrown out when he'd landed his ship.

"She's a little big to miss isn't she?" Toby asked, moving toward the second fighter, leaving the last one to land for K'narf.

"She'll be cloaked when we approach her." Peter said, climbing up toward his rear cockpit.

"Cloaked, you've cloaked that whole ship?" K'narf asked, remembering just how big that thing had been.

"And applied stealth technology to her too, she's neigh onto invisible to radar as well." Peter said, settling into his seat and strapping himself in.

"This I've got to see." K'narf said, quickly climbing the ladder that had been thrown out for him and settling into what he had thought of as the pilots seat; apparently it could be flown from either cockpit he realized as the cockpit sealed itself and they very gently lifted off.

"At least tell us when we're heading in her direction." Toby flashed, using the common frequency so all of them could hear him.

"If she was a snake she'd bite you." Peter flashed, sending them a picture of a coiled rattlesnake preparing to strike.

"I think I see her." K'Narf flashed, as he began to see a hazy outline in the stars they were flying toward. "She's dead in front of us but I can't tell how close." He added, as he confirmed in his own mind that the blurry outline in front of them was Big Mama, as they called her.

"Yeah, I think I see her too." Both Toby and Karl flashed at the same time just as Brent de-cloaked her.

Dear sweet Jesus, K'Narf thought to himself, as a huge monster of a ship suddenly appeared in front of them. A ship that bore no resemblance to the one he and T.J. had attacked. "What in the world did you do to her?"

"We gave her angles." Peter flashed, as he stopped their forward motion. "No matter what direction a radar beam hits her it will be shattered and deflected making her as close to invisible as possible, especially when you throw in the cloaking paint."

"How do you get in her?" Toby asked, flashing the question that had just occurred to both

K'Narf and Karl.

"From the bottom, that's why they parked those two landing craft outside the ship when they set her down on Pluto; once down they couldn't get their ships in or out of her."

"Landing craft, why did you call them landing craft?" K'Narf asked, picking up on the new term for the Snark scout ships.

"Because she was originally designed as a troop transport, and what you thought were scout ships were actually landing craft for troops." Peter flashed, sending them pictures of how fully equipped fighting men could be packed into the Snark landing craft like sardines and dropped down to whatever planet they were targeting. "They could drop fifty men at a time onto a planet and round robin back for more in an awfully short time." He added, almost sounding like he was bragging on the Snark military.

"Troop transport, how many troops could she hold?" Karl asked, flashing his question as Peter moved them into a position directly under her, followed immediately by the fighters carrying K'Narf and Toby.

"She has quarters for five thousand fighting men. Four men to a compartment with each compartment having it's own power supply, emergency water and food in case of an attack." Peter answered, as he lifted them up through a force field and into Big Mama's belly.

"Five thousand?" Toby exclaimed, knowing how deadly a force of five thousand fighting men could be.

"And her crew?" K'Narf asked, flashing his question as lights were suddenly turned on and lit up a massive landing bay.

"Up to a thousand from the looks of the crews quarters." Peter said, maneuvering his fighter to a landing cradle and setting her down, quickly followed by K'Narf and Toby's fighters in the adjoining cradles. "But we only found seventy three of them when we cleaned her out."

"What did you do with them?" K'Narf asked, flashing his question, knowing Peter meant dead bodies.

"Sent them to the sun, same place they sent that Helper's city." Peter flashed, sending them a mental image of a large container being launched at the sun, presumably full of Snark bodies.

"How many fighters can you park in here?" Karl flashed, changing the subject as he looked at the rows of fighter cradles circling the huge bay.

"We tore out some of the adjoining bulkheads so we could accommodate all three hundred fighters and those two little beasts over there."

"What did you do to them?" K'Narf asked, looking over at what had been the two Snark scout ships, wondering as he looked at them which one had almost killed him.

"Made them invisible. They're our new launch platforms for the paintball missiles and for dropping off the black widows."

"What's a black widow?" K'Narf asked, turning his attention back to Peter.

"Nasty little critters that can track a warm breathing body through six inches of steel and put a one-eighth inch laser beam right through its heart. We want to salvage those warships, and Brent did not want them damaged any more than necessary, so we designed the black widow to specifically hunt down and kill our little angels as they like to call themselves." Peter said, as he climbed out and closed his fighter's hatch. "Satan's little black hearted angels if you ask me." He added, as he dropped down to a gangway just below his fighters cradle.

"Sounds like a deadly weapon." Toby said, mentally envisioning a black widow spider as he unstrapped himself.

"They are, that's why we only made thirty six of them, three for each Snark warship."

"You think three will be enough?" Karl asked, climbing out behind Peter.

"Yes, more than enough and be careful getting out, you'll be in freefall until they put a spin back on her."

"You put a spin on her?" K'Narf asked, sticking his head out of his cockpit as he asked the question.

"Yes, it keeps stuff from floating off desks and tables, but not much else. Less than half of the gravity you have in the Whale." Peter said, grabbing at a gangway support as Big Mama suddenly started to move out from under him. "We always stop her when we have someone coming or going. It's safer than trying to land and take off with her moving."

"Why would they be using a troop transport as a mining ship?" Toby asked, climbing out of his cockpit.

"Best we could figure out was that she was mothballed. It seems our eager beaver miners leased her for twenty percent of the take according to the captain's records."

"Twenty percent of what?" K'Narf asked, jumping over to the same gangway Peter had landed on.

"You asked me that once before, so why don't you let me show you?" Brent asked, suddenly dropping down, seemingly out of nowhere, and joining Peter and K'Narf.

"Lead the way, and thank you Peter." K'Narf said, as Toby and Karl both jumped over to their gangway and joined them. "And your two buddies." He added, as he turned and waved at the other two pilots who were sailing across the landing bay on a trajectory leading to a portal on the far side.

"You're more than welcome." Peter said, leaping at the same portal his pilots had jumped at. "Anytime!" He shouted back, leaving them in the care of Brent.

"So, what was so valuable that it was worth their time to come all the way out here to get it?" K'Narf asked, turning to face Brent as he asked his question.

"Follow me." Brent said, turning and heading for an open portal at the end of the gangway. Then led them, barely touching the floor, as they pulled themselves along using handrails in a mutual agreed upon silence, through what seemed to be at least a mile of interconnected hallways. Finally they came to a heavy steel door with a combination lock welded into it where the door handle should have been.

"What is this, a vault?" Karl asked, checking out the workmanship.

"Exactly what it is." Brent said, as he started to punch buttons. "Here we go." He added, as the door swung open to his touch. "Go ahead, take a look."

"What is it?" K'Narf asked, looking in at what

appeared to be a large pile of reddish looking stone-like crystals.

"Well, it seems that red diamonds are as rare in their System as they are here."

"Those are all red diamonds?" Toby asked, looking at the size of some of the crystals and instantly realizing the value of what he was looking at. "What weight are we looking at?" He added, stepping into the room and reaching down and picking up a rather large specimen.

"Don't know about carats, but physically you're looking at a little over three hundred pounds."

"You've got to be kidding!" Toby exclaimed, holding his specimen up to the light. "Are you going to try and sell'em?"

"If we can find a place where we can phony up a mine without the government stepping in and taking it away from us." Brent said, as he looked over at K'Narf. "Well, what do you think of our treasure?"

"I know they're valuable, but it seems like an awful lot of trouble to go to for some pretty red rocks." K'Narf said, thinking of the blood that had already been shed over them, and the blood that was yet to be shed.

"Well, now that you've all been duly impressed, let me show you the rest of the place." Brent said, as Toby threw his fist size diamond in the rough back onto the pile and Brent shooed them out and relocked the door.

"Why do you keep it locked?" K'Narf asked, wondering who Brent didn't trust.

"Don't know really, just always lock it back the way I found it."

"How did you find it, the combination I mean?" Karl asked, thinking it a reasonable question.

"Captains log, found a picture of you in it too." Brent said, turning to look at K'Narf.

"Me!" K'Narf exclaimed. "What do you mean me?"

"Seems they got a pretty good shot of you when you flew over them in that van of yours. Your headlights blurred it a little, but you can definitely make you out through the windshield." Brent said, turning around to look at K'Narf. "Found a picture of Wonder's little Volkswagen in the scout ship that fired on her too." He added, turning back around and heading down the hallway before K'Narf could see his face.

"What are these?" Karl asked, sticking his head into a large cubicle with four fold-up beds chained to the wall."

"That's a four man cubicle, in the back you'll see a bathroom, shower, and a small kitchenette where they could cook if they wanted to or had to if the ship had been damaged." Brent said, stopping as he stuck his head in the cubicle. There's a total of twelve hundred and fifty of these. Each one of them had it's own reserve water and food supply along with an atomic battery for power. This is where we've been getting some of our batteries for the vans and the fighters."

"I was wondering about that." K'Narf said, sticking his head into the cubicle for a look-see.

"Yes, I bet you have." Brent said, backing out so Karl could get a look inside.

"How many have you scavenged so far?" K'Narf asked, turning his attention back to Brent.

"Not as many as you might think. There's one in every cubicle, but we found almost a hundred in their bulk storage. We also found much larger spares for this ship and some almost as big for the scout ships. Etaguf City is running on one of the larger spares right now." Brent said, as they all continued to pull themselves along the hallway, only occasionally touching the hallway floor.

"Where are you taking us now?" Toby asked, curiosity getting the best of him as the hallway they were traveling through seemed to be endless.

"Our greenhouse." Brent said, as he slowed down and stopped at a hallway that had turned into a glass-bottomed walkway. The floor was lit up with extremely bright lights and they could all see green plants just starting to grow in what appeared to be long stainless steel garden beds filled with black looking earth. A color that told K'Narf it was extremely rich soil. "We fired her back up just a little over two weeks ago so the seeds we planted are just starting to grow." Brent added, looking down at the endless rows of planted beds.

"Rudy would love to see this." K'Narf said, thinking of his Head Gardner.

"Who do you think did most of this?" Brent asked, a look of mild surprise on his face as he turned to look at K'Narf. "When Rudy found out about this place, he hot-footed it straight over here. He was the one that made us clean out all of the dead plants, then found their seed storage and showed us how to replant everything. When he left here he had a big box of seeds and was in such a hurry to get back to plant them that he asked me for a ride back in one of my fighters."

"Did you give it to him?" K'Narf asked,

reminded again just how far out-of-the-loop he had allowed himself to get. He was beginning to wonder if his personally helping Mitch fix that line had really been that important, or if he had only been stroking his own ego.

"Of course, he should be planting them right about now. He was like a kid at Christmas, dozens of new vegetables but mostly he was excited about some new oxygen producing plants that had apparently been designed to grow in a near weightless environment." Brent said, turning as he continued across the glass-bottomed walkway.

"That should make his green zones over in Etaguf City a lot more interesting." Karl said, knowing how Rudy had been more than a little worried about what to plant.

"I confess, I'm a little surprised, I thought that at least you would have kept yourself updated where it concerned Rudy." Brent said, stopping as he turned to look at K'Narf.

"I'm a little surprised that my Head of Security didn't deem it necessary to tell me that a man had been killed up here." K'Narf retaliated, looking back at Brent.

"Are you *suggesting* that it's part of my job description to keep you informed about what's happening, even if it's something not directly under my control?" Brent asked, stressing the word *suggesting* as if it was a swear word.

"Yes, I believe that's exactly what I'm asking." K'Narf said, starting to get angry.

"So, if it's just a *suggestion*, I can decide whether or not to follow it based on how I might, or might not feel at the moment, is that correct?" Brent asked, looking at K'Narf in a bemused fashion, no sign of anger or emotion showing on his face as he spoke.

"Are you *suggesting* that I make my request a direct order?" K'Narf asked, stressing the same word as he realized that there was no anger in Brent's face. He's testing me, he thought as he added. "If you have a Bible handy I could have you swear on it."

"A direct order I know how to handle." Brent said, letting a slight grin show on his face. "And you don't need a Bible, you have my word. Just remember that once an order is given it's almost never possible to retract it." He added, obviously quoting someone since the word structure he used was not his own.

"I smell T.J. in this." K'Narf said, seeing Brent's eyes light up at the mention of T.J.'s name.

"If T.J. had lived during the time of Daniel, she would have been known as his equal in wisdom." Brent said, openly smiling now that his and T.J.'s point had been made. Sometimes it was necessary to give orders rather than make suggestions and K'Narf needed to learn this.

"How big is this thing?" Toby asked, trying to change the subject as he looked down at the greenhouse stretching out below his feet, realizing as he did that K'Narf and Brent had just had one of those small life-changing moments.

"If you put all the levels together it would cover about three acres. Not exactly what you would call small in anyone's book." Brent said, turning and walking away as the bottom of the walkway turned back into steel decking."

"Where are you taking us now?" K'Narf asked, struggling just a little to keep up with Brent, seasoned as he was to traveling Big Mama's hallways.

"You told me to turn this ship into a state-of-the-art command ship, so I'm taking you to the command center that I believe has fulfilled that *order*." Brent said, stressing the word *order* as they entered what was obviously a very large dining area.

Just as big as Redwing's Commissary." K'Narf thought, except that the tables were bolted down and very utilitarian, plus the fact that there were barely a dozen people in the place. Looking at the small portion of the serving line that had been activated, the food choices seemed to be limited too.

"I don't think Redwing has anything to worry about." Toby said, looking over at the diners as they saw them and waved. "At least not from a competitive point of view anyway." He added, taking a dig at Brent.

"Same food." Brent said, waving back at his crew just as they entered another long hallway that seemed to go on forever.

"Holy Moses!" Was all K'Narf could say as they finally entered a huge room that by it's shape was the forward nose section of the ship. It had more television screens in it than K'Narf had ever seen in one place.

"Before you say anything let me give you a tour." Brent said, motioning for them to follow him down a set of steps onto what appeared to be a raised dais with six large comfortable looking chairs, with the first three chairs on a slightly higher level than the second three chairs. "These are the command chairs. The three chairs on this lower level will be occupied by myself here, Mister Zebulon over there and whoever is 'it' sitting there in the center." He added, stressing the word "it" as he winked at Toby.

"What do you mean by 'it'?" K'narf asked, Toby's expression telling him that he had just triggered a verbal trap.

"Well, it seems we're getting ready to play a serious game of hide and seek with our Snark friends, and in every game of hide and seek I've ever played someone always had to be it, and since that's your chair I guess that means you're it!" Brent said, grinning from ear to ear.

Ah, the teenage Brent I know how to deal with, K'narf thought, seeing that both Toby and Karl had joined Brent in gleefully watching for his reaction. "I don't understand, what's this hide and seek thing you speak of?" K'Narf asked, playing dumb. "Is this another one of those movie things?" He asked, turning to look at Toby as he watched them all do a double take, thinking it impossible that anyone could grow up and not know about playing hide and seek. A game his brother had taught him well in an abusive sort of way.

"Aw come on!" Brent exclaimed, looking at Toby and Karl for support. "Don't tell me you've never played hide and seek?"

"Do I still get to sit in the center chair if I say yes?" K'Narf asked, watching their faces as they all realized he had just turned the tables on them.

"Score one for K'Narf." Toby said, looking at the surprised expression on Brent's face as K'Narf walked over and sat down in the front and center chair.

"What, no buttons for me to push?" K'Narf asked, as Brent regained his composure. His well rehearsed dig having gone sadly astray.

"No, whoever's *'it'* doesn't get to do anything but watch." Brent said impetuously, then grinned a little sheepishly as he walked over and sat down in the chair on the far side of K'Narf. "Karl why don't you go ahead and set down in one of the VIP chairs." He added, as he pointed at one of the three chairs behind them on the second level.

"That's an awful lot of screens." Toby said, looking at a bank of thirty-two large video screens arranged in horizontal rows of eight stacked four high forming what amounted to a large movie screen.

"Please notice that there are fifteen positions on both your left and right for a total of thirty operators. Each operator will be monitoring ten fighters with their screens divided into ten segments, nine of which will be reduced to a viewing area of three by five inches with the tenth one taking up the remainder of the screen. By a simple click of their cursor they can change their selection, and whatever fighter's camera they select is what we will be seeing on the corresponding screen in front of us. We will be seeing exactly what the pilot of that fighter is seeing through his windshield!" Brent exclaimed, repeating himself in an effort to make sure K'Narf and Toby understood what he had just said.

"There are thirty-two screens in front of us." K'Narf pointed out.

"Yes sir, the top left and top right screens are what the pilots flying our two modified scouts will be seeing. We will be able to witness the firing of the paintball missiles, and the subsequent placing of the black widows as well." Brent said, pointing at the two screens in question.

"How about audio?"

"The fighter to fighter chatter will be piped into this room for all of us to hear. Dan Ripplinger and Peter Vanlieu are my flight leaders, so most of that chatter should come from them, although all of my pilots can pitch in at any time including the pilots of the two scout ships. I'd like to also point out that everything you see and hear will also be recorded for later evaluation. This will be the most documented fight in the history of warfare!" Brent exclaimed, waving his hand in a half circle that took in all the various positions and screens in the room. "Including any word or expression you might choose to express." He added, pointing at a small, six-inch recording ball similar to the one used by Minh Le at their council meeting that was hanging just above and slightly in front of where they were sitting.

"Are we being recorded now?" Toby asked, feeling slightly over whelmed by the total picture of what Brent had presented.

"No, you're safe to say anything you want." Brent said, looking around K'Narf at Toby. "At the moment anyway."

"It's far and away more than anything I could have imagined. Saying you did an outstanding job is by anyone's standard an understatement. I don't know what else to say." K'Narf said, thoroughly impressed at what Brent had put together.

"You made me swear on my word of honor that I would not get personally involved in this fight if you allowed me to take out the Snarks that killed Wonder. Well, this is the closest I can come to being in the coming battle without being physically involved." Brent said, reminding K'Narf of the oath he had taken.

"And these three VIP chairs." Karl asked, kind of hoping he might be one of them.

"They are reserved for three very special guests." Brent said, seeing the disappointment on Karl's face as he realized he was not going to be one of them.

"There's going to be a gallery up there." Brent said, pointing at a landing up and behind where they were sitting. "That's where any of my crew not directly involved will be sitting. You're more than welcome to join them." He added, seeing a smile wipe away the disappointment that had been on Karl's face just seconds before.

"And these three special guests, who are they?" K'Narf asked, knowing the answer but wanting to hear Brent say it.

"I can't tell you that yet." Brent said, as he stood up.

"What about I *'have your word'* about keeping me informed?" K'Narf asked, reminding Brent of his promise as he stood up and looked down at the diminutive figure that was his Head of Security.

"I can't tell you something that I don't yet know." Brent said, looking up at K'Narf as Toby and Karl stood up with them.

"You're sure enough to prepare seats for them." K'Narf pointed out.

"Yes, that much I can tell you." Brent said, giving K'narf a cryptic smile.

Chapter Fifteen

"K'Narf, where are you?"

"In my office, why." K'Narf flashed back, knowing it was Brent.

"We need to get together, where can I meet you?"

"How about here in my office?"

"No, it has to be somewhere where we won't be disturbed. How about meeting me over at Minh Le's where we can hole up in one of his classrooms?" Brent asked, sounding excited.

"Right now?"

"Yes, right now if that's possible."

"Okay, give me about ten minutes, and I'll meet you over there." K'Narf flashed, ending the connection. It must have something to do with my setting down with a fresh cup of coffee and a sweet roll, he thought, standing up as he took a sip, thinking he could eat and drink as he walked.

It had been a little over three months since he had ordered Brent to keep him updated, and his earnest desire to fulfill that order was starting to drive him nuts. Not that he really wanted him to stop, he thought, there was no way he could keep up with what was going on without Brent's constant updates though at times, like right now, they tended to be a nuisance.

T.J. was definitely showing now, and she was having some unusual mood swings, but nothing really out of the ordinary if what Wanda was telling him was true.

The fighter production was on schedule and they were happily building an apartment building up at Etaguf City, if that's what you could call it since it was in actuality a freestanding life support unit. In fact, everything is going along so well, I wonder what it is that has Brent so excited that he has to tell me in person.

"Minh Le, how are you?" K'Narf asked, walking into the front lobby of Minh Le's school and running into the headmaster himself.

"I am just fine, Brent asked me to meet you and show you what room he has selected." Minh Le said, smiling as he took the hand K'Narf had offered and shook it. "Right this way please." He added, as he turned and headed down a central hallway that led into the main building. Then, stopping at an unmarked door, he took out a set of keys and unlocked and opened it revealing a stairway lit only by the light coming from the hallway they were standing in; a stairway that led down to what was obviously a rarely used basement.

"Is Brent down there?"

"Not yet, but he is on his way. He asked for this room so the two of you would not to be disturbed. There is a small table and two chairs down there, but you might want to dust them off before using them." Minh Le said, like it was the most natural thing in the world to be offering him this particular room.

"Where's the light switch?"

"Oh, I am so sorry, here." Minh Le said, reaching just inside the door and flipping a switch that instantly lit up the stairway and the room at the base of it. "Please." He added, motioning for K'narf to enter. Meaning he has other things to do than stand here and hold a door open for me, K'Narf thought, as he started down the stairs, feeling trapped as the door closed behind him.

There was, as Minh Le had said, a small table and two chairs in the room and they really did need a dusting before use. There were also several large cabinets against one wall and some odd bits and pieces of lumber stacked against another. The ceiling was mostly pipes going in several different directions, at least none of them showed signs of leaking, he thought, as he picked up a chair and tried unsuccessfully to shake the dust off of it. Seeing some rags hanging out of one of the cabinets he took the liberty of borrowing one; giving the table and both chairs a good wiping down just as he heard the door at the head of the stairs open.

"K'Narf!" Brent yelled, as he charged down the stairs. "Ah, there you are!" He added, as he hit the bottom of the stairs and spotted K'Narf, then hit him with a poke at his brain that almost knocked him to the ground.

"Oww, that hurt!" K'Narf said, staggering back a full step. "What do you think you're doing, have you lost your mind?"

"Keeping a promise." Brent said, grinning.

"A promise to who?" K'Narf cried out, backing away from the grinning Brent.

"You, you said you wanted to know who was going to be setting in those VIP seats, didn't you.

then protect yourself!" Brent said, flashing K'Narf a detailed picture of what he had to do to protect his brain from another attack and then hit him again.

"Oww!" K'Narf yelled, as his instincts kicked in and formed a protective shield around his brain just as another mental bolt hit him. "What in the..."

"Good, now the next time you're hit, try and backtrack and find out who it is that's trying to probe you." Brent said, again flashing K'Narf detailed instructions on how to do just that.

"I know exactly who just hit me, I'm looking at him!"

"Are you sure?" Brent asked, walking over and pulling out a chair and setting down as K'Narf was hit by another hard driving probe.

"Hey, that wasn't you!" K'Narf said, this time staggering all the way back into the lumber stacked against the wall behind him.

"No, it wasn't. Now, like I said, next time he strikes, try and tell me who it is that's trying to probe you." Brent said, leaning back in his chair and smiling just as K'Narf was hit again.

"Someone named Addi!" K'Narf exclaimed, realizing that he had just backtracked the source of his attacker and caught the name of Addi before he was blocked.

"Good, now come on over here and sit down and make that protective shield permanent!" Brent ordered, again flashing K'Narf specific instructions.

"What exactly just happened?" K'Narf asked, as he followed Brent's instructions to the letter and sealed and locked a protective shield around his brain with a ten digit password making it permanent unless he personally removed it.

"You can't probe someone with your shield

up which makes an attacker vulnerable to backtracking if his or her shield is down." Brent said, leaning forward and putting his elbows on the table and looking directly into K'Narf's eyes as he sat down opposite him. "You now know how to protect yourself, and how to identify an attacker if someone should try and probe you. You might say you have an extremely high level security clearance now, and it would be safe for someone with a similar clearance to contact you if it became necessary."

"What are you trying to say?"

"Someone wants to talk to you." Brent said, you can contact him on this frequency." He added, sending K'Narf an extremely high number.

"Who?" K'Narf asked, as he mentally dialed the number, knowing it was someone named Addi almost before he asked the question. A fact that was instantly confirmed by a flash from a source he had never encountered before.

"Mister Etaguf, my name is Addi, and I'm sorry we had to be introduced this way, but extreme caution has been our way of life for many centuries now."

"Addi, uh, and you are?"

"I am one of the gentlemen your Mister Conrad happened upon in St. Catherine's monastery out in the Sinai. I believe, if we understand him correctly, that you sent him out here to look for us. Is this not so?"

"Yes, uh, yes, it is." K'Narf stammered, suddenly realizing that he was communicating with a man that had known the Lord personally. "Uh, how can I help you?" He flashed, not knowing what else to say.

"Actually we thought maybe there was an

area or two that we might be of some service to you." Addi flashed. "But first there is a small problem of security."

"Security?"

"Yes, security. If one of those Snark probes, as you call them, should happen to probe one of your people they would know instantly all about your organization and thereby all about us. This we would not like, as they might have time to pass that information back to their superiors before your Mister Conrad has a chance to destroy them. This would not be good either for you or for us."

"Those probes can read minds?" K'narf asked, seeing the surprised look on Brent's face when he mentioned the probes, having spoken as well as flashed his words.

"Yes, even though the odds are greatly against them accidentally probing one of your people, they do have that ability in a mechanical sort of way."

"What do you suggest?"

"Mister Conrad has assured us that you possess the authority to require all of your people to learn to protect their minds as you have learned to do. Is he correct?"

"Yes, I do have that authority." K'Narf said, knowing that he did indeed have that authority or status as T.J. called it.

"Then please use it. When all your people are protected, please contact me on this frequency. We would like to attend one of your Council meetings in this place you call the Whale, if that would be satisfactory to you?" Addi asked, sending his message in as soft a way as was possible when flashing.

"We?"

"Yes, all three of us. Like I said, there are a few areas that we may be of service and we would like to have the opportunity to present them to your staff, with your permission or course."

"I cannot think of anything that would please us more." K'Narf flashed, not knowing what else to say.

"Thank you for your kind words, then until I hear from you, and salutations to your T.J., my brother and I are much looking forward to meeting her." Addi flashed, cutting the connection.

"What's that bit about those probes being able to read minds?" Brent asked, as soon as he realized that K'Narf was no longer communicating with Addi."

"That's what he said, why?"

"Had something in mind. Good thing I waited." Brent said, as he changed the subject. "Well, what did he want?"

"They want to meet with us." K'Narf said, briefly wondering at Brent's cryptic comment as he answered him.

"When?"

"Just as soon as we can get everyone protected."

"That's what I thought he wanted." Brent said, thinking of the questions he had been asked after they had taught him to protect himself. "How do you want to go about this?"

"We each teach someone, and that someone teaches someone else. Everyone becomes a teacher. Simple exponential math should handle the rest. I suggest you start with T.J. since you and her are so close." K'Narf said, trying to look as if what he had just said made perfect sense.

"You want me to go slam T.J. with a probe that will cause her to talk to me in Mandarin?" Brent asked, knowing K'Narf would know exactly what he was referring to when he mentioned Mandarin. "Suicidal I am not, besides, I made Toby a promise that I'd start with him." He added, grinning.

"Promise me that you will be no harder on him than you were on me, meaning that you will send him the same specific instructions you sent me in the same order and timing." K'Narf said, knowing how easy it would be to do a few extra probes before sending the antidotal instructions.

"You do know how to take the fun out of life."

"Promise?"

"Yes, I promise."

"In fact, make that an oath everyone takes so no one can use this as a way of torturing their buddies, okay?"

"Is that an order?"

"Yes, you have a problem with it?"

"No, it's just that it's so much easier to pass on an order than a suggestion." Brent said, grinning in a way that told K'Narf he would do as asked.

"How long do you think it will take?" K'Narf asked, thinking this would not be easy.

"We have a lot of people that will have to be trained, not counting my search teams that are scattered all around the globe. They'll have to be called in as well. My best guess is a minimum of two months." Brent said, trying to mentally add up all the people that will have to be hunted down and trained.

"Thought it might be something like that." K'Narf said, frowning slightly as he realized he would have to wait at least that long before he could find out what Addi meant by an *'area or two'* that they might be of service. I just pray that one of those *'areas'*, K'Narf thought, stressing the word *'areas'* in his mind, is some detailed knowledge of that fleet heading at us and the resources behind it.

"By the way, Karl wants you and T.J. to come out to Etaguf City for a ribbon cutting ceremony next week." Brent said, changing the subject.

"Ribbon cutting?"

"Yeah, they're finishing up their first apartment building and they want you and T.J. to cut the ribbon, one of the perks of being *'it'*." Brent said, laughing at the look on K'Narf's face. "Now I need to go find Toby before anyone gets a chance to warn him." He added, his meaning clear as he pushed back his chair and stood up.

Chapter Sixteen

"Karl." K'Narf said, greeting Karl as he opened Lady's door and got out.

"Welcome back." Karl said, stepping back as K'Narf headed around to the other side of the Lady.

"Sorry it took us so long to get here, but in T.J.'s condition we couldn't travel very fast." K'Narf said, as he opened T.J.'s door. "Oh, and Wanda came along with us, she wouldn't let T.J. out of her sight, and she wanted to find out what that errant better half of hers has been up to while she was at it." He added, helping T.J. out and smiling at the face Wanda made at him when he called Rudy her better half. Remembering at the same time how his grandfather used to call himself that when he wanted to aggravate his grandmother.

"Who's holding down the fort back at the Whale?" Karl asked, realizing that both Rudy and Wanda were now in Etaguf City.

"Edna, Edna Kennedy, new lady that agreed to take Redwing's place. Tough little lady, just wait until you meet her." K'Narf said, as he helped Wanda out.

"She's a nice lady!" Both T.J. and Wanda said at the same time. "Don't you go listening to Mister negative here." Wanda added, giving K'Narf an eyeball-rolling look as she scolded him.

"Yeah, as long as you do exactly what she tells you!" K'Narf exclaimed, getting a nasty look from T.J. this time.

"She's just a little strict, someone has to be with all those newbies coming and going all the time." Wanda said, defending Edna's behavior. "And besides Thomas and Judy both like her."

"That's because she feeds'em." K'Narf said, turning to Karl. "Sorry about the little family dust-up. You had something to show us?"

"Yes, we've completed our first apartment building and we wanted to have an official ribbon cutting ceremony and we can't do that without you two." Karl said, beaming at K'Narf and T.J. as he ushered them toward the stairway that led from the sixth level where they had landed, down toward the first where they could see a tall structure looking like something out of a Mediterranean setting with its stone and brick facade and general coloring.

"What's this all about?" T.J. asked, realizing that there were people lined up on both sides of the wide walkway all the way down to the building they were heading for; people that were now standing up and starting to applaud.

"I afraid you two are kind of like royalty to them. They all know that they wouldn't be here if it hadn't been for the two of you and this is the only way they can show their appreciation. I'm afraid you'll just have to suffer through it. I did order all of them to stay back, they're not allowed to approach closer than ten feet or they'll get trench digging duty."

"Trench digging duty?" K'Narf asked, taking

T.J. protectively by the arm as they moved forward.

"Digging foundation trenches, they have to be nice and deep and they have to be dug out with pick axe and shovel. Roughest duty I have, so I use it to threaten them with." Karl said, as he ushered them in between the two rows of applauding people, some with tears in their eyes as they looked first at T.J. and then at K'Narf, alternately curtsying and then bowing as they walked past them.

"Maybe those history readers weren't such a good idea." K'Narf said, thinking maybe he had made a mistake insisting on them.

"No!" Karl exclaimed. "Those orientation readers pulled us together and made all this possible. We wouldn't be here ready to fight if it weren't for those readers, especially the one where we were all forced to make a choice of standing or running away like the space people did. It meant as much to me personally." He added, as he stopped and turned to face K'Narf and T.J. "As swearing 'Never Again!' does to a new Israeli officer when he takes his oath of allegiance."

"But we..."

"We're not trying to claim that you are our King and Queen, or even that you're an elected President and first lady, but we believe in you and what you are trying to do and will follow you into hell itself if that's what it takes." Karl said, cutting K'Narf off as he started walking again, shooing a few people away that had started to crowd in when they had stopped walking.

"This is a beautiful walkway, what kind of stone is this?" T.J. asked, deftly changing the conversation away from her and K'Narf.

"We used marble from Brent's Marble

Mountain for this particular walkway." Karl said, happy that T.J. had changed the subject. "Each walkway has been created with a different stone, red rock from Arizona and obsidian from Hawaii for example. That way, once you've memorized the different materials used, you will know exactly where you are in the City by the color and texture of the walkway you're using." Karl said, swinging his arm in a circle that took in the whole city as he gave his explanation.

"And these planters?" Wanda asked, joining in the conversation as she pointed to the huge thirty-five foot wide planters that paralleled the walkway on both sides with their banks of grow lights covering them some forty-five feet overhead, lights that also lit up the walkways all around the City.

"Rudy had us build those, they're part of his green zones. Made us dig them out so they'd be deep enough too." Karl said, remembering the argument he had lost about that very subject.

"Dig them out?" K'Narf asked, wondering why they would have had to be dug out.

"We wanted the sides of the planters to be sitting height. I mean just high enough to be comfortable to sit down on if you wanted to stop and take a rest, and that didn't make them deep enough for what Rudy wanted, so he made us dig them deeper."

"How much deeper?" Wanda asked, knowing Rudy had a will of iron when he chose to use it.

"Every one of these planters, and they parallel every walkway in the city, are a full seven feet in depth and filled with some of the best black delta bottom loam we could find, every ounce of it pesticide free too." Karl said, obviously proud of what they had done, even if it was under duress. "And all the watering systems are in as well. All we're waiting on now is the seedlings Rudy said he has growing." Karl said, looking over at Wanda like maybe she knew when they would be ready.

"Don't look at me, I haven't seen that man in weeks." Wanda said, defending herself. "That's part of the reason I'm up here." She added, looking around as if Rudy might pop out of the crowd at any minute.

"You want to find Rudy, just climb up in one of those planters and start tromping around, he'll be down on you in a matter of minutes." Karl said, daring Wanda to do just that.

"I'll take your word for it." Wanda said, as they started down the steps that led to level three.

"Wanda!"

"Speaking of you-know-who." Karl said, turning around and seeing Rudy coming up behind them. "Where are those seedlings you promised us?" He asked, knowing his constantly asking about them got Rudy's goat.

"Go ahead, keep asking and I'm going to take that ten gallon hat of yours and use it for a planter." Rudy said, smiling from ear-to-ear as he grabbed Wanda and lifted her off her feet.

"K'Narf, T.J., 'bout time you two dropped in. You ain't gonna believe what's been going on around here." Rudy said, letting go of Wanda long enough to give K'Narf and T.J. both a quick hug.

"So we've been hearing." K'Narf said, flashing T.J. privately that he had never seen Rudy so animated before.

"He's found a new purpose for his life." T.J.

flashed back, just as they reached the bottom of the stairs and started walking toward the level two stairs.

"Did Brent tell you about the seeds they found on that ship of his?" Rudy asked, hugging on Wanda as they walked.

"Yes, he did say something about you're getting some seeds. Did any of them sprout?"

"Yes, all of them did. I planted them in one of the gardens down in the park and they're growing like crazy. I've got thousands of the little oxygen seedlings coming up." Rudy said, looking over at Karl as if daring him to say something. "And some really weird looking vegetables, gonna drive the cooks loopy trying to figure out what to do with them though." He added, giving Wanda another hug.

"How do you know they're oxygen producing plants?" T.J. asked, getting interested in what Rudy was saying.

"Read about them in the gardener's log over on Brent's ship. He knew if anything ever happened he would have to replant. That's why he had so many seeds in his reserve stock. Had enough on hand to replant the ship and start a little over two thousand of them over here."

"Speaking of oxygen, you seem to have a perfect mix but your not rotating like the Whale, and I don's see any circulating fans, how do you do it?" K'Narf asked, then added. "And heating, this place feels like a near perfect seventy degrees, how do you do that too?"

"Both real simple. This old planet still has some heat in her belly and we tapped into it with heavy gage copper cable."

"Copper cable?"

"Yeah, we drilled a one point seven inch hole a little over two thousand feet deep underneath each pillar and stuffed a heavy gage copper cable down into it, then we attached it to the base when it was installed. If you go over and touch the steel bases of one of the pillars you will find that it is radiating a temperature of one hundred and three degrees. Now that heat is transmitted up through each section of the pillar until it reaches the top of the column. Top of a tall column registers about sixty and a short one about ninety. That heat differential is what keeps the air stirred up." Karl said, talking as though he was getting tired of explaining this particular facet of the city.

"Of course, heat rises and cold air descends. What about your oxygen and nitrogen mix?" K'Narf asked, understanding now how the city kept its air stirred and heated itself.

"Every pillar segment is setup as a dispenser. It has sensors that constantly measure the air and adds oxygen or nitrogen as needed. Like I said, it's almost too simple."

"Then when Rudy's seedlings grow up it will cut down dramatically on your need to replenish oxygen." K'Narf said, mostly for his own benefit.

"Seedlings, seedlings, I've been hearing about these things for weeks and I still haven't got anything for my guys to plant." Karl said, taking another dig at Rudy.

"You know what, friend or no friend, next time DeeDee asks me to help her out with a little something, I'm gonna do it!" Rudy exclaimed, getting a smile from everyone except Karl.

"Aw come on, I haven't done anything that bad." Karl said, thinking Rudy might be serious.

"Gonna lay off me about them seedlings?"

"If you'll lay off me about that hat they bought me."

"Deal!" Rudy said, reaching over and shaking Karl's hand.

"Did they really buy you a ten gallon hat?" K'Narf asked, remembering DeeDee's remark about his going to need one.

"Let's just say it's a large cowboy hat, and yes, it got here a couple of days after you left." Karl said, looking over at K'Narf and then at T.J., knowing she had had something to do with it. "Anyway, here we are." He added, as they came up to an area in front of the new apartment building that had been roped off.

"You were really serious about that cutting a ribbon stuff weren't you." K'Narf said, seeing a bright red ribbon stretching across the odd looking doorway. "It looks a little massive." He added, noticing how thick the walls seemed to be where the strange looking doors had been installed.

"She is!" Karl exclaimed, leaning back as he looked up at her. "Built her right up into the roof so she'd be extra support in case anything ever happened. Gonna build all of them that way."

"Your apartment building is a roof support?" T.J. asked, wanting to know if she had heard right.

"Yes, her frame is made of heavy gage I-beams and they are pressed right up against the steel of the roof. When all of the buildings are built, and this little fight we're about to get into is over, we're going to add about fifty more feet of dirt and rock to the top of the dome. Deep enough to take a direct meteor hit as long as it wasn't some kind of monstrous asteroid." Karl said, bragging a little as he answered T.J.'s question.

"And those odd looking doors?" K'Narf asked, looking again at the oddness of the doors.

"Let's make this official and cut this here ribbon and I'll show'em to you." Karl said, as one of his men walked over and handed him an oversized pair of scissors. "Would you do the honors?" He asked, as he handed the scissors to T.J.

"Beautiful!" Karl exclaimed, as everyone started applauding as T.J. cut the ribbon. "Here let me have those." He added, as he took the scissors out of T.J.'s hand and handed them back to the man who had given them to him.

Turning around, T.J. watched as most of the people started leaving and heading back up towards level six where all their tents were. I wonder how many of them missed their sleeping hours just to come and see this and us she wondered, as she turned her attention back to Karl and his new building.

"What you're looking at is two pressure sensitive doors that recess back into the wall with a small one person airlock in the center. The doors will always remain open unless there is a drop in pressure that would indicate a major rupture somewhere in the city.

"A rupture?" Wanda asked, instantly thinking about Rudy's safety.

"Yes, if something happened like a meteor hit that penetrated or created a crack that allowed our atmosphere to start venting, these doors would automatically close and seal themselves up against that center airlock. The airlock, of course, would make it possible for emergency responders to get out of or into the building." Karl said, answering

Wanda's question. "All of our buildings will be designed like this." He added, as he led the way into the building proper.

"The doors are designed to remain open?" T.J. asked, looking at the edges of what apparently were massive recessed doors in the thick walls as they followed Karl into the building.

"Yes, this main hallway runs straight through the building on this level to an opposing set of doors built into the opposite wall. Each level is designed and built the same way. Notice that the hallway is slightly curved. When the adjoining apartment building is built this hallway will connect with it and so on until eventually the connected hallways will form a complete circle intersected only by the walkways." Karl explained, as he pointed out the slight curvature when looking down the softly lit hallway.

"You say every level is built the same way?" K'Narf asked, mentally picturing a connected hallway circling all the way around on each level of the building.

"Yes, this city is designed for walking, and speaking of walking I would like to direct your attention to our large bay windows. We didn't want this to turn into an area where no one knew who their neighbors were, so one of the fellows suggested we install big bay windows." Karl said, walking over to the nearest window and looking in on an empty room. "They all look in on the living room and can be closed off simply by drawing the drapes during your sleep period or if you simply want some privacy."

"Why would you want to do that?" K'Narf asked, not sure of the reasoning behind the big bay windows.

"I know you aren't old enough." Karl said, answering K'Narf. "But you two remember back before we had television, don't you?" Karl asked, making it sound more like a statement than a question as he looked at T.J. and Wanda.

T.J., prompted by Karl's question, turned and looked at Wanda with a slightly raised eyebrow, then both of them turned and looked at Karl, neither of them offering assent or dissent.

"Okay, okay." Karl said, realizing that neither of the girls were going to answer him. "Anyway back before television, everyone used to sit out on their front porch and talk to their neighbors as they walked by. That's the intent here. We don't have television and we don't want our people to become closed off, so this is what we came up with.

Every apartment, be it an efficiency unit designed for a single person, or a larger one, two or three bedroom unit, has a large bay window looking in on the living room." He added, looking at T.J. and Wanda to see if they approved.

"I think it's a wonderful idea." Wanda said, remembering the many pleasant evenings she had spent on her grandparent's front porch. "They can leave their front door open and say hi as people walk by." She added, seeing how it could work if they left their front door open.

"Yes, I would agree with Wanda, it is a wonderful idea." T.J. said, which settled the issue as far as Karl was concerned. Not that K'Narf's opinion didn't matter, but in this situation it just didn't matter that much.

"Well, I like your building, but there are a few things that I need to talk to the two of you about, and if I'm right it will be Sabbath in less than two hours so we don't have a lot of time. T.J. and I brought sleeping bags and we can stay in the Lady, and I assume that Rudy has a tent somewhere. I just hope it's big enough to accommodate Wanda." K'Narf said, seeing a look in Karl and Rudy's eyes that told him instantly that they were up to something. "We'd like to get a bite to eat up at Redwings before we settle in too, if that's possible." He added, keeping a wary eye on Karl and Rudy as they looked at each other.

"Well, you just going to stand there or are you going to show'em?" Rudy asked, looking at Karl.

"Show'em what?" K'Narf asked, knowing he was right about Rudy and Karl hiding something.

"Ladies, if you would come this way please." Karl said, totally ignoring K'Narf's question as he took off walking just a few steps ahead of them until they reached an apartment that had it's bay window curtain drawn.

"And this is?" K'Narf asked, not liking the feeling of being ignored.

"This is your apartment!" Karl exclaimed, opening the door to the smell of cooking. "Please." He said, stepping aside and letting T.J. and Wanda in followed by K'narf and Rudy into a fully furnished living room.

"T.J., Wanda!" Redwing cried out, as she came out of the kitchen. "Your early, we're not ready yet."

"Ready?" Wanda asked, sniffing at the smells coming from the kitchen.

"Ten minutes, I'll have everything on the table in ten minutes." Redwing answered, rushing over and hugging T.J. and then Wanda.

"I thought you didn't have television up here." K'Narf said, looking at the large flat screen mounted on the wall.

"We don't, but we have VHS, DVD and Blu Ray players. There is a lot of good Christian programming that we have hauled up here." Karl said, pointing at a small stack of DVD's on a table next to the couch. "And you wouldn't believe what they've rigged up in some of those tents up there." He added, pointing in the general direction of the sixth level.

"T.J., Wanda, Redwing said we have ten minutes. That's more than enough time. You two take Redwing into the kitchen and work her over in there." K'Narf said, motioning for T.J. and Wanda to take Redwing back into the kitchen. "These two birds belong to me." He added, as he turned to look at Karl and Rudy.

"What do you mean, '*work her over*'?" Rudy asked, looking as though he was about ready to make a run for it as he repeated K'Narf's words.

"Please gentlemen, take a seat and make yourself comfortable." K'Narf said, settling down on the couch and stretching out his legs.

"Not until you tell me what you mean by '*work her over*'." Rudy said, edging toward the still open door as he repeated K'Narf's words a second time.

"It'll only hurt for a second or two, now don't be such a chicken and set down, both of you." K'Narf ordered, watching as they both sat down on the front edge of the two visitor chairs that had been placed opposite the couch.

"What do you mean '*it'll only hurt for a second or two*'?" Karl asked, repeating K'Narf's words as he picked up on Rudy's nervousness.

"Just that!" K'Narf said, as he hit them both with a hard driving probe.

"Oww, that hurt!" They both yelled at the same time.

"Then protect yourselves!" K'Narf said, as he flashed them a detailed picture of what they had to do to protect their brains from another attack and then hit them both again.

"Oww!" They both yelled, as their instincts kicked in and formed a protective shield around their brains just as another mental bolt hit them.

"Good, now make that protective shield permanent!" K'Narf ordered, flashing them specific instructions on how to do just that as he heard Redwing scream from the kitchen.

"What just happened?" Rudy asked, as he followed K'Narf's instructions to the letter and sealed and locked a protective shield around his brain with a password that only he would ever know.

"It seems that those two Snark probes running around out there have the ability to probe a human mind, and if they should happen upon one of our people the game would be over before it started."

"Why can't some of Brent's fighters just blow them out of the sky?"

"Losing one probe is acceptable, that's why they sent three, but losing all of them would be like sending them a message that we're waiting for them." K'Narf said, setting up and staring at the two of them for a few seconds to make sure they knew that he was serious about what he was going to say next. "I'm sorry I had to hurt you, but it's the fastest way to convince your brain that it has to protect itself, and you have to leave here and start doing exactly what I did to you to everyone up here. Everyone becomes a teacher until there are none of our people left unprotected. This is going to be a lifelong protection that will not be removed except by death." K'Narf said, letting them both know just how critical the situation was.

"Yes sir." Rudy and Karl both said in unison.

"Oh, and tell them only one hit and then complete instructions just like I gave you, no trying to hit their buddies multiple times before telling them how to make it permanent. Tell them that's an order, okay?" K'Narf asked, not really expecting an answer.

"Karl." T.J. said, sticking her head out of the kitchen. "Would you like to stay for dinner, we have plenty and you would be more than welcome."

"Uh, I'm uh, sorry, I uh, have already made other arrangements." Karl stammered, as he got to his feet. "Oh, by the way, there's a fully furnished guest bedroom for Wanda and it's big enough for two people if she wants to share it with him, I guess." He added, as he pointed at Rudy.

"You've got a dinner date with DeeDee don't you?" Rudy asked, knowing he was right by the look on Karl's face. "Well I'll be bit by a rattlesnake if'un I ain't right." Rudy said, grinning from ear-to-ear.

"K'Narf." Karl said, stepping over to where K'Narf was sitting. I brought this with me thinking you and T.J. might enjoy it during your Sabbath down time. I found it a couple of months ago on Amazon." He added, taking a DVD out of his inner coat pocket.

"What is it?" K'Narf asked, taking the offered

DVD.

"It's called 'The Feasts of the Lord" by a pastor named Mark Biltz.

"And you liked it?"

"Yes, very much, and I think you and T.J. will too. Anyway, it's yours to keep if you like it, I can get another copy sent up." Karl said, as he backed away toward the still open front door.

"Give my best to DeeDee!" Rudy yelled, as Karl left and headed down the hallway.

"Something else you need to tell me?" Rudy asked, thinking there was something that K'Narf had left out of his little speech.

"Not until our next council meeting, and we can't do that until everyone is protected, and I mean everyone." K'Narf said, looking down at the DVD in his hand.

Chapter Seventeen

"Well, should I contact *him*?" K'Narf asked, stressing the word "*him*" when he looked at Toby and Brent, both of them knowing he meant Addi.

"Best I can tell we're ready." Toby said, looking at Brent to see if he agreed.

"It took a lot longer that I thought." Brent said, thinking of the three months it had taken to track down and train some twenty six thousand glow heads and teach them to protect themselves. "But we've checked and double checked everybody. Even made a contest out of it by offering an all expenses paid week anywhere they wanted to go if they could find someone that was unprotected, and it's been over two weeks since we've had a winner."

"You're both saying we're ready?" K'Narf asked, wanting to hear both of them actually say it.

"Yes." Brent said, with Toby nodding his head in agreement.

"Okay, here goes." K'Narf said, as he mentally dialed the frequency that Addi had given him.

"Ah, Mister Etaguf, we have been waiting anxiously to hear from you." Addi flashed, instantly responding to K'Narf's open connection.

"Yes, well it took a little longer than we

initially thought, but we believe we're ready if you are." K'Narf flashed, speaking his words as he flashed them so Toby and Brent could hear what he was saying, and making sure that Addi knew that he wasn't alone in his assessment that they were ready by saying "we". "When would you like to meet with us?" He added, thinking the sooner the better.

"Could your Mister Conrad pick us up at the main door to St. Catherine's Sunday night at midnight. There is no moon that night and with that cloaking thing he does with his van we should be safe from prying eyes." Addi flashed, sending K'Narf an image of three men in monk's habits along with several trunks and suitcases waiting in front of an ancient stonewall.

"You're bringing luggage with you?" K'Narf asked, surprised at the image Addi had sent.

"Yes, we must. We have been here many centuries and experience has taught us that when one leaves, even for a short time, anything he leaves behind must later be retrieved at a considerable cost of energy and time." Addi flashed, letting K'Narf know that anything they left behind would not be there when they returned. "We hope to be as little inconvenient on you as possible." He added, in a soft almost humble tone.

"I assure you, that you and you're friends will be our most honored guest as long as you want to stay with us, and Sunday night is excellent as that will give me time to round everyone up and get them to the Whale." K'Narf flashed, feeling very pleased with himself as the date was finally set to meet with Addi and his friends; friends he knew to be survivors of that original attack and a direct source of information on the enemy fleet that was now just few months away.

"It will be our deepest pleasure, and I pray that you will never know what it is like to be isolated from those of your own kind for centuries on end as we have been." Addi flashed, giving K'Narf a glimpse into his soul as he added. "Until Sunday night then." And broke the connection.

"I heard you say 'Sunday night'." Brent said, quoting K'Narf.

"Yes, they want you to pick them up Sunday at midnight just outside St. Catherine's." K'narf said, flashing Brent a copy of the picture Addi had sent him.

"So that's why you said they were bringing luggage." Toby said, as K'Narf sent him the same picture

"Hey, I know that wall, it's just outside the main gate." Brent said, noticing the stonework in the background. "Good, there's plenty of room there to drop the van down." He added, mentally making plans to do just that.

"Monday in the Whale then?"

"Yes, Monday." Toby and Brent both agreed.

"Thomas!" K'Narf yelled, as Thomas landed on his stomach and then walked up and laid down next to his head and started purring at a level that literally vibrated the bed.

"Does that cat know that it took us over twelve hours to get here." T.J. moaned, referring to the fact that they had to travel extremely slow because of her pregnancy, which was now definitely starting to show, as she tried to roll over and cover her head.

"Yes Thomas, I love you too." K'Narf said, as Thomas started licking at his face.

"Judy!" T.J. exclaimed, feeling rather than seeing the second Manx cat land on the bed and then snuggle up to where her face was hidden beneath the covers and start purring.

"It seems we're supposed to get up." K'Narf said, scooting back and sitting up.

"Edna, have these cats been fed?" T.J. flashed, her head still under the covers.

"Yes ma'am, over an hour ago." Edna flashed back.

"Then why are they over here bothering us?" T.J. flashed, uncovering her head as she reached over and rubbed the back of Judy's head.

"Maybe 'cause they're your cats and they ain't seen the two of you for some time now!" Edna snapped back.

"Ain't our fault they refuse to live anywhere but here, and who's here and who ain't anyway?" K'Narf flashed, asking Edna for an impromptu head count, having seen a lot of vehicles on the landing pad when they came in last night, but because it was during a sleep cycle he hadn't had a chance to ask.

"Everyone coming from that city of yours got here Friday evening just before Sabbath started except them pilot team people. Seems they can fly a lot faster than everyone else so they didn't leave there until Sabbath was over."

"You mean Dan and Peter?" K'Narf asked, thinking of Brent's two lead pilots.

"Yeah, that's them, and their copilot bondies." Edna flashed, referring to Dan and Peter's bond mates, Susan and Antoinette. "Most everyone else is here." She added, not sounding too happy about it.

"Where's here?" K'Narf asked, wondering where that many people could be hanging out in the

Whale.

"Most of'em are still down in the vans they came in 'cept for Wanda and Redwing who're in here messing around in my kitchen!" She exclaimed, obviously upset by whatever they were doing.

"Rudy's here too?" K'Narf asked, wanting to make sure Rudy had left his precious plants long enough to join them for today's council session.

"Yeah, but he ain't out'ta bed yet. Kicked poor old Ralph out and made him go sleep down in the garden in a sleeping bag!"

"Who's Ralph?"

"Ralph's our new gardener, somebody had to take over the garden now that Rudy's abandoned it. Been sleeping in their room since they ain't been using it, 'cept now of course." Edna flashed, thinking someone like K'Narf should know these things.

"Has anyone heard from Brent?"

"Yeah, he's still about an hour out, and who're these mystery people he's bringing in here with him anyway?" Edna asked, having overheard Wanda and Redwing talking about them as they were unloading some of those fancy plates and flatware they'd brought with them.

"You'll have to come to our council meeting to find out." K'Narf flashed, thinking it only polite to include Edna since she was here. "And bring your Ralph in with you."

"First thing, he ain't my Ralph, not by a long shot! Second it ain't none of my business, so if you don't mind I'll just take myself into my bedroom and get a little rest since it's clear that I'm not needed in the kitchen!" Edna flashed, obviously upset by Wanda and Redwing taking over her kitchen as she broke the connection.

"What was that all about?" T.J. asked, struggling to sit up in the waterbed.

"Edna's just being Edna." K'Narf said, throwing back the covers and sitting up on the padded bedrail. "But apparently Brent's just an hour out and I'd like a few minutes in the Chapel before they get here." He added as he stood up, staggering a little at the much lighter gravity in the Whale.

"Help me out of this thing before you go." T.J. said, struggling to roll her body around so she could get her legs over the edge of the bed.

"Aren't you getting close to being due?" K'Narf asked, knowing it had been almost eight months since that night she had taken him out to dinner.

"Yes, best that Wanda and I can figure is that I'm due now, but the baby hasn't dropped yet and it still seem to be too small."

"Too small?"

"Yes, we both think that I should be much larger by now, unless it's a very small baby." T.J. said, struggling up onto her feet with K'Narf's help. "You go on now, I can make it from here. I'll meet up with you in the conference room." She added, not wanting K'Narf to feel how worried she was about the size of their baby.

"Good." K'Narf said, stepping into the chapel and seeing that it was empty.

Walking up the center aisle, he dropped to his knees in front of the softly lit crystal cross and folded his hands in his lap. "Dear Lord, well, this is one of the moments I prayed for. I asked for knowledge and wisdom concerning who or what it is that we're up against in this fight I've got everyone involved in, and with Your blessing I will have those answers in the next few hours.

I'm not sure what, if anything, I need to ask for now except the wisdom to understand whatever it is they tell us, and that we will all have favor with them. I guess I'm here mostly to say thank you, not just for sending Addi and his two friends to us, but for the child T.J. is carrying and for what has been happening out on Pluto as well as here in the Whale and back at Sutton Wells.

Everything seems to be going along far beyond anything I could have imagined or planned, and that is because You are personally involved with everything we are doing. So I ask again, please be in that conference room with us today, and when it's all over and we leave that room, let any and all decisions and actions decided on be in accordance with Your will. That is my prayer, thank you for listening. We'll talk again tonight when this is all over, I love you, Amen." K'Narf said, as he finished praying and stood.

Seconds later, using the yellow guy rope to pull himself along, he stopped and looked at the landing pad with its jumble of parked vehicles. There was of course the Lady in her customary position, but scattered around her was a hodgepodge of vans mixed with two state-of-the-art mirrored Teardrop fighters that had just landed. Guess I'd better git down there and get this circus organized before they get here, K'Narf thought as he grabbed the guy rope and started pulling himself along, letting go only when he had gathered enough speed to fly along with it.

"Make a hole!" K'Narf yelled, using a term Rudy had taught him as he realized that the incoming pilots were still standing in the landing spot in the kitchen.

"Gotcha." Dan said, as he grabbed K'Narf and pulled him back down to the floor.

"Thank you, one of these days I'm going to learn how to land without bouncing off the floor."

"I doubt that!" Rudy said, sitting at the kitchen table with a big grin on his face.

"Wanda, you're looking great. Rudy said you'd put on a few pounds, but I don't see it." K'Narf said, smiling at Wanda as he headed toward the conference room entrance.

"Lord have mercy, if that boy had black on him, I'd slap it off!" Rudy said, grabbing his coffee mug and getting up. "I didn't say any such thing!" He added, looking at Wanda and taking off after K'Narf.

"You're lucky there's witnesses!" Rudy exclaimed, entering the conference room and seeing K'Narf already seated next to T.J.

"What did you do to Rudy?" T.J. asked, knowing K'Narf had done something.

"Let's just say my Yin and Yang are in harmony." K'Narf said, smiling at Rudy as he sat down on the other side of the table.

"If I get bruised up, you get bruised up, you just remember that." Rudy said, pointing at K'Narf with his coffee mug.

"That's enough, both of you just stop it!" T.J. said, hitting K'Narf on the shoulder to make sure he was listening to her.

"Okay, okay." K'Narf said, looking around the room. Knowing they were all smiling because of the little show he and Rudy had just put on. "Debbie, haven't seen you in a long time. That mural you did in the Commissary is really something." He added, noticing that Debbie looked much more mature. No longer the little girl, K'Narf thought, nodding at the pilots Dan, Susan, Peter, Antoinette along with Hacker, all sitting up on the back row in the exact same seats they had used last time, their dark skin and golden glowing eyes in stark contrast to Hacker and everyone else in the room.

"Thank you." Debbie said, visibly pleased with being singled out and with the compliment.

"Minh Le, I see your ready to record everything." K'Narf said, seeing the silver ball sitting on it's pedestal in the center of the table.

"Well, looks like we're all here." Toby said, entering with Irena on his arm followed by Premo and Sylvia then by Karl with a blushing DeeDee on his arm. "Oh, and just in case you didn't know it Karl and DeeDee are our newest bonded couple." He added, standing back and applauding as DeeDee blushed and Karl took an exaggerated bow.

"Congratulations!" K'Narf said, standing up and joining in the applause, glancing down at the obvious joy on T.J.'s face, knowing she was taking the lion's share of credit for DeeDee's success. "Wait, before you sit down, we need to pull three seats up here to the front. Rudy you'll have to get up and scoot over one seat." He added, trying to make sure their guests would have a front row seat.

"Rudy, you need to save three seats next to you, not just one for Wanda." T.J. said, seeing Rudy pulling over a seat next to him for Wanda.

"Why?" Rudy asked, looking confused.

"Brent and Redwing will need a place to sit too." T.J. said, ignoring the look K'Narf gave her as she helped him arrange the seating, knowing he knew that she was still playing matchmaker.

"Speaking of Wanda and Redwing, where are they?"

"Getting ready to bring in fresh baked cinnamon rolls and hot coffee, where else." Toby said, pulling out a seat for Irena at the back end of the conference table. "I can smell them from all the way back here." He added, letting K'Narf know that he wasn't overly pleased at losing his front place at the table.

"Edna said I was supposed to go to a meeting, is this where I'm supposed to come?"

"You must be Ralph." K'Narf said, remembering the invite he had offered to Edna and the new gardener. "Please come in and take a seat."

"Yes, please do, there's an empty one right over there next to Debbie." T.J. said, pointing at an empty chair, smiling as she helped K'Narf out with the seating arrangements a second time.

"Thank you ma'am." Ralph said, quickly moving over next to Debbie and taking his seat.

"When was the last time you heard from Brent?" K'Narf asked, looking at Toby as he sat down.

"Two minutes ago, they're in the airlock now."

Chapter Eighteen

"Ladies and gentlemen." Brent said, as he entered the conference room and stepped aside, allowing the three men with him to enter, each of them wearing a simple monk's frock with the hood thrown back. "Please welcome Addi, his brother Zithri and their nephew Neli."

"You are most welcome." K'Narf said, standing up. "I'm K'Narf Etaguf and this is my wife T.J., and you are?" K'Narf asked, pointing at T.J. and then offering his hand.

"I'm Addi." Addi said, taking K'Narf's hand and shaking it. "And this is my brother Zithri and our nephew Neli." Addi said, as K'Narf shook each man's hand in turn, noticing as he did that there was a distinct sense of age in Addi and Zithri that was not evident in Neli.

"I'm not real sure how to start this off." K'Narf said, stepping back and looking at the three men just as Wanda and Redwing came in carrying trays loaded down with hot cinnamon rolls and coffee. "Okay, let's get you seated and go from there." He added, as he pointed out the chairs they had reserved for them

"Do you gentlemen drink coffee, if not I have

hot tee, canned orange juice and water as well." Wanda said, setting a tray loaded with coffee mugs and two full carafes of coffee down on the table in front of them. "And Redwing has hot cinnamon rolls fresh out of the oven to go with whatever you choose." She added, as Redwing set her tray down and left immediately to go get another tray followed out of the room by Brent.

"Coffee would be just fine for Zithri and I, though Neli should probably have water." Addi said, pulling out his seat and setting down.

"I too would prefer coffee." Neli said, ignoring his uncles as he took the chair between them and sat down.

"Very good." Wanda said, as she set coffee mugs, creamer, sugar and a carafe down in front of them and let them serve themselves as she pushed the still loaded tray over in front of K'Narf and T.J. "Now, how about a sweet roll to go with that coffee?" She asked, not waiting for an answer as she placed a beautiful little gold trimmed plate with a still hot cinnamon roll on it along with a set of napkin wrapped fancy flatware in front of each of them just as Redwing and Brent returned carrying two more fully loaded trays into the room.

"I'll take care of my pilots." Brent said, seeing Hacker sitting on the second level.

"Oh, I don't mind." Redwing said, stepping around him and heading for the back of the room.

"Okay, is everyone set?" K'Narf asked minutes later, having given everyone time to enjoy their sweet roll and coffee before asking his question.

"Ask if one of them would like to act as spokes person for the others." T.J. Flashed, trying to help.

"Well, I guess it's question and answer time." K'Narf said, standing up. "Addi, would one of you like to act as spokes person, or do you prefer to respond individually?" He asked, trying to follow T.J.'s advice.

"Yes, though Neli would prefer to answer all your questions, we won't let him, so it would be best if we individually answered any questions you may have." Addi said, noticing that T.J. caught the twinkle in his eye.

"He's picking on Neli!" T.J. flashed, surprised as she tried to let K'Narf know what was going on.

"Uh, then uh, good." K'Narf stammered, confused for a second by T.J.'s flash. "Then I uh, get to ask the first question. If we defeat this fleet coming at us, what can we expect them to hit us with next. I mean what are their resources, like how many fleets like this one do they have and how many warships will they send against us next time?" He added, struggling to get his question out, even though it had been on his mind for a long time.

"You do not ask the simplest of questions first, do you." Zithri said, as he looked over at his brother. "Do you want to answer this or shall I?"

"Could you speak up please?" Dan asked, from the back of the room.

"I'm sorry, it may be best if I stand." Zithri said, having received a "you take it" nod from his brother. "How's this?" He asked, as he stood up and stepped over to the head of the table.

"Yes, that's better, thank you." Dan said, as the other pilots and Hacker nodded their heads in agreement with him.

"First off let me say that my brother and I are

not historians by training or vocation but I will do the best I can. Now, how do I answer your question without regaling you with stories of suffering and death, because most of our history revolves around those particular subjects." Zithri said, sadness touching his face as he continued.

"Many, many thousands of years ago God created three beautiful planets very similar to your Earth. They were all located well inside what your scientists call the goldilocks zone around a star almost identical to your sun, and they each possessed all of the factors required for beautiful and highly habitable planets including moons to both stabilize their rotations and to stir and circulate their oceans.

With the passage of time, and the advent of space travel, they were all colonized and trade was established between the flourishing economies of the various nation groups, and between the governments and businesses of the different worlds. Then something strange happened on the world of Carthy."

"Is it necessary to speak of him?" Addi asked, interrupting his brother with his question.

"It is *he* they will be fighting when that battle group gets here, so *he* must be mentioned. How else can I answer the question that has been asked?" Zithri asked, looking at his brother, stressing the word "*he*" as he spoke. "Maybe if I used an analogy rather than our actual history it would be easier to explain." He added, still looking at his brother.

"Yes, I believe that would be more appropriate and understandable, if there is such a word as understandable that can be attributed to him and his Carthy." Addi said, nodding at his brother to continue.

"Sorry, we are just trying to figure out how to best explain what happened on our home worlds and how what happened there is going to affect you here." Zithri said, taking a deep breath and exhaling as he continued. "On the planet called Carthy a new leader rose up and viciously subdued all of the nations and peoples of his home world. He was, using the best example I can think of, the equivalent of the leader your history books call Hitler, only on his world he won. Declaring himself to be a god, he had his storm troopers quickly do the same to the worlds of Leecan and Arctan. The carnage exceeded that of your holocaust."

"Didn't anybody try and fight him?" Toby asked, seeing images of the holocaust in his mind.

"God did not give the peoples of our worlds the gift of anger that he has given you!" Addi exclaimed, standing and speaking for his brother. Those that refused him obeisance were led to the slaughter like sheep and those that survived now live the life of slaves, bought and sold in markets similar to those of ancient Rome. You are a fortunate people, you did not suffer in silence you struck back, something we did not know how to do." Addi said, his voice dropping as he sat back down, giving the floor back to his brother.

"However, all did not go his way." Zithri said, taking up the story where Addi had left off. "Those that lived and worked in space defied his rule and fled and he lives in fear of them to this day. That is why his fleet is coming this way."

"Why, we are not space people." K'Narf said, trying to understand as he looked first at Zithri and then at Addi.

"His probes have seen your international space station. To him that is a threat, anyone in space is a threat. He knows that the people cannot rise up as long as he controls space, and knowing that there are hundreds of thousands of people, most of them from his own world living somewhere out in space, terrifies him."

"And this fleet that is coming at us are his equivalent of Hitler's storm troopers?" Toby asked, beginning to understand what they were up against.

"Yes, the fleet coming at you is the best he has. It was designed for one purpose and that is to seek out and annihilate the space people. He is constantly sending out probes, like the ones he sent here, looking for any evidence of their existence, and like I said your international space station is to him a major threat." Zithri said, looking first at Toby and then at K'Narf.

"And if we destroy it, this elite fleet of his, what will he send at us next?" K'Narf asked, getting to the answer he wanted most to hear.

"Nothing, he maintains only three other fleets, one for each of the planets under his control, but these fleets are not warships of the quality and scale of those coming at you. They are mostly designed to drop troops down to the planets surface if and when they are needed, and that has not happened for thousands of years. At least that is what Addi and I believe since that is how long we have been marooned here with no one coming to look for us."

"Nothing." K'Narf repeated softly, feeling a heavy load lift from his shoulders as he repeated the word. "Nothing, you're sure about that?"

"Yes, other than probes, and he will send

probes, probably lots of them, and you should do your best to destroy them before they can report back." Addi said, speaking up again for his brother.

"Brent?" K'Narf asked, letting his question hang in the air as he looked at Brent.

"It will not be difficult to detect and destroy them when the time comes, and I thank you for your information." Brent said, giving both Addi and Zithri a slight bow of respect. "Now my pilots and I will feel much easier about the job in front of us, knowing as we now do who and what it is that we will be destroying." He added, as he turned and looked at Dan and Peter, his flight leaders, as they both gave him a nod of understanding.

"Is there anything else you can tell us that might help?" K'Narf asked, looking back at Zithri.

"We had planned to tell you the direction that they would be coming in from, but your Mister Conrad has already figured that out." Zithri said, nodding at Brent. "It seems he has been tracking the probes when they go out to the edge of your solar system and has marked the direction they fire their monthly update missiles at."

"We applaud his ingenuity." Addi said, adding to what his brother had said.

"It seemed the logical thing to do." Brent said, looking a little embarrassed at the sudden praise.

"Anything else?" K'Narf asked, thinking there ought to be more.

"Are you aware that they are barely forty light years away?" Zithri asked, thinking how close the triad planet system actually was might be of value.

"No, but thank you for telling me, knowing

this will help me calculate travel times."

"Is there anything else?" K'Narf asked, looking at Zithri and then at Addi.

"We would also like to offer our services during the actual battle."

"What service?" Brent asked, surprised by the offer.

"We should be able to locate and listen in on their command frequency and their fighter to fighter chatter. We know the dialect they will be speaking and most of the slang they will be using if it hasn't changed too much. We would like to think it may be of some value if we can tell you what they are saying and planning before they do it." Addi said, speaking again for his brother.

"That is an offer we will gladly accept, and we would also like to take you out to Brent's command ship and let you critique what he and his people have done and how he has planned the battle?" K'Narf said, seeing Brent nod his approval as he made the offer.

"We would be most happy to look at what Mister Conrad has prepared, but you must understand that our vocation does not even remotely resemble any kind of military training."

"What exactly is your vocation?" T.J. asked, jumping into the conversation now that it was no longer limited to a military subject.

"We are very happy you asked us this question." Addi said, as both he and Zithri smiled at T.J. "We are early cycle mentors, and Zithri specializes in prenatal mentoring." He added, nodding at his brother.

"Prenatal and early cycle, what do you mean?" T.J. asked, trying to make sense out of what they were saying.

"Think of us as specialized teachers. We both have your equivalent of a PhD in pre-birth through the fifth year cycle of male child rearing. We survived the attack because Neli here was having a temper tantrum and we had taken him to the other side of the planet where he couldn't bother anyone when it happened." Zithri said, grinning at Neli as he said it.

"Tell them how old I was!" Neli exclaimed, speaking up for the first time. "You'll have to forgive my uncles, I was only three years old at the time." He added, deciding to defend himself.

"You were only three years old." T.J. said, understanding now why Neli had not joined in on his uncle's presentation; looking over at Wanda who was following the conversation as keenly as she was.

"Their skills and training only go through to the fifth year or cycle as they like to say." Neli said, looking over at T.J. "So from then on I pretty much had to raise myself." He added, getting a smile from T.J. for his effort at being humorous.

"We apologize for our great, great nephew, it is sadly apparent that proper adulthood mentoring has been lacking, especially in regards to showing respect for his elders." Addi said, scolding Neli with words to go with the look he was getting from his uncle Zithri.

"It's seven 'greats' uncle Addi." Neli said, trying to get in a last little dig at his uncles.

"Two are sufficient to convey meaning, and I'm sure they are aware that we are considerably older than you." Zithri said, the tone of his voice letting everyone in the room know that he was not happy with his nephew's attitude in front of strangers.

"Exactly how old are you?" K'Narf asked, getting a slight dig in the ribs from T.J. as he asked the question. "If you don't mind my asking." He added, not knowing why T.J. was upset.

"We are approaching twenty, what in your culture you would call middle age." Zithri said, looking at Neli, who was having as sudden coughing fit, with a look that was easily readable by everyone in the room. "You have something you would like to say nephew?" He added, offering the floor to Neli.

"No, no, it's just that I've never thought of the two of you as being middle aged, it always seemed to me to be somewhat greater than that, that's all." Neli said, smiling at his uncles as he got control of his cough and leaned back in his chair.

"By twenty, you mean twenty thousand?" K'Narf asked, making sure he understood right as he got another dig in the ribs from T.J.

"Yes, and as our nephew has noted our skills have not been used for a long time, but I'm sure they will come back to us as the need arises." Addi said, looking at T.J. and K'Narf. "And we would very much like to offer them to you and to your expectant child." He added, as Zithri nodded his head in full agreement.

"You said 'male child rearing', what did you mean by that?" Wanda asked, having followed the conversation very carefully.

"Our training was specifically designed around the development of the male child, we have never trained for or attempted to mentor a female."

"Why, what's the difference?"

"Well, not to put too fine a point on it as your culture likes to say, the emotional range of a male is straight forward and easily manageable. Their training is not overly difficult as evidenced by Neli here; however, a female is, uh, how should I say this.

"Devious, isn't that the word you like to use uncle Zithri?" Neli asked, smiling over at T.J. with a mischievous twinkle in his eye.

"No, no, it's more like they have a much broader range of emotions and they learn to use them at a very early stage in their development." Zithri said, trying to ignore Neli's comment. "There are those that specialize in mentoring them but we, my brother and I, have never attempted a female as yet."

"But you do know how, don't you?" Wanda asked, looking at Zithri for her answer.

"We have taken the courses, so yes, we do know the theories and techniques involved."

"Then you could do it?" Wanda asked, pressing the point.

"No, but we would be willing to teach someone else how to do it." Zithri said, looking pointedly at Wanda.

"Then why do you offer your services to me since it is obvious that I am going to have a baby girl." T.J. asked, her hands feeling the smallness of her stomach and the baby in it; a baby that still had not dropped even though it was now several days past due.

"On the contrary T.J. ma'am, you have a very healthy boy developing in you. I have already felt his presence." Zithri said, smiling at T.J. and K'Narf.

"How can you say that when I am so small?" T.J. asked, looking at Zithri in confusion.

"Brother." Addi said, holding his hand up as a signal to his brother to let him talk. "T.J. ma'am, are you believing that the gestation period for your baby is to be nine months?"

"Yes, of course I am, why?"

"Please forgive us, my brother and I take some things for granted."

"Like what?"

"Like thinking that you knew the gestation period for the child you are developing within you is thirteen months, not nine." Addi said, seeing the instant shock on T.J.'s face.

"What, oh no!" Sylvia cried out, instantly causing every eye in the room to turn and look at her.

"Is there something you need to tell me?" Premo asked, turning in his chair so he could look Sylvia in the eye.

"No, uh, I mean not yet. I mean I'm only a few days late, that's all!" Sylvia exclaimed. "Now stop looking at me like that, I'm not pregnant.

"That settles it, we are moving up to Etaguf city. I am not carrying this baby for another five months down in that gravity hole at Sutton Wells. Karl, is that apartment you let us use still available?" T.J. asked, looking down the table at Karl.

"The three bedroom, yes, it belongs to the two of you unless you would like one on a higher level."

"No, that one will do just fine, is there another one nearby that our friends here could stay in?" T.J. asked, pointing at Addi and Zithri.

"Yes, there is an empty two bedroom just across the hall that would be perfect for them." Karl said, looking at Addi and Zithri.

"Finally!" Neli said, grinning from ear-to-ear. "They'll finally have someone to occupy themselves with and practice on besides me. You think you could find me a small efficiency or even an empty tent. I would be most happy with either." He added, still grinning from ear-to-ear as he looked over at Karl.

"I'm sure that can be arranged." Karl said, looking at the smiling Neli.

"You will sing a different tune when your first son is due to be born and you are desperate to find mentors for him." Zithri said, trying to counter Neli's glee at getting away from them.

"Neli is going to be a father?" T.J. asked, surprised by Zithri's revelation.

"No." Addi said, correcting what Zithri had just said. "But now that he is among his own kind it should not take long for a pretty girl to catch his heart."

"Oh, I see." T.J. said, realizing that the young Neli was an available bachelor. "We have many young girls that would love to meet him, don't we Debbie?" She added, as she leaned forward and looked down the table at Debbie.

"Stop it T.J.!" Debbie said, blushing.

"And you sirs." T.J. said, turning to look at Zithri and Addi. "Is there an opportunity for your hearts to be caught?" She asked, feeling emboldened by the way the conversation had turned to the subject of matchmaking.

"No, the ones we loved were in that city you saw pushed into the sun. There will never be another for us, and now at our age it would be like bonding with a child; a child that would not reach middle age before they lost us. No, I'm afraid our time of bonding has long since passed."

"Is there anything else that you would like to

speak about before we adjourn this meeting?" K'Narf asked, realizing that both Addi and Zithri seemed to be about talked out.

"We do have one more area where we might be of some little assistance." Zithri said, motioning for his brother to stand as he pulled out his seat and sat down.

"Uh, yes." Addi said, as he stood up and looked around the room until his eyes rested on Rudy. "I understand Mister Marcus that you are the person that loves plants." He added, looking at Rudy.

"Yes, uh, I mean I do." Rudy stammered, surprised at being singled out.

"I too love plants and have knowledge of most of the plants that grow on our worlds. I understand from Mister Conrad that you have acquired seeds for many of them, is that correct?"

"Yes, I have planted most of them over on that ship of Brents and the rest of them in the gardens at Etaguf city, why?"

"I would like to offer my assistance to you as some of them are temperamental and difficult to get started. I also have many recipes in my mind that your cooks might find helpful as well." Addi said, smiling at Rudy. "I consider my self an amateur cook as well as a gardener." He added, nodding to K'Narf as he sat down.

"If it will keep Rudy happy and off our backs we will all be grateful to you." Karl said, jumping at the chance to give Rudy a dig. "And I flashed my people and they will have that apartment of yours fully furnished by the time we get back. We also have small efficiencies available on literally every floor. You can take your pick when we get there."

He added, looking at Neli.

"Addi, have you ever looked at plant auras? We can you know, we can look at them using my Kirlian glasses. I'll show you when we get back." Rudy said, excited now that he had someone that loved plants like he did to talk with, ignoring Karl as he leaned out over the table in order to see Addi's face.

"Is there any other business we need to deal with at this time?" K'Narf asked, trying to regain control of the meeting. "Good, then we have a lot to do. Minh Le, see that everyone gets a copy of this meeting and make sure it's added to the orientation readers."

"Yes sir, uh, unedited sir?" Minh Le asked, not sure T.J. would want her personal information included.

"Yes, unedited." K'Narf said, seeing the reluctant agreement in T.J.'s eyes. "Now let's have a moment of silent prayer before we adjourn."

Chapter Nineteen

"Are you about ready?" K'Narf asked, for the third time.

"Yes, and if you ask one more time I am going to find a way to hurt you." T.J. said, picking up her new car coat and putting it on over her space suit as she looked around the room. All of their personal things were already packed in the Lady, and all of the heavier things were to be taken up on one of Karl's big rigs. He had promised her that they would not only beat her there, but that he would have them arranged in their apartment when they got there. It would be nice to have her own things around her up there instead of the furniture Karl and his buddies had picked out.

"Just leave the key in the lock, Karl's guys will need to get in when they come to get the heavy stuff." K'Narf said, seeing T.J. hesitate and then put the key back in the lock.

"It's just that we were happy here, and I've not had much of this kind of happiness in my life. I just hate to leave it. I really thought that I was going to have our baby here." T.J. said wistfully, as she reluctantly turned and headed for the Lady and what she knew was going to be a long trip, especially knowing what she was going to be in for wearing her newly grown and fitted space suit with it's unique disposal abilities.

"We don't have to go. You can still change your mind." K'Narf said, knowing that moving to Etaguf city had been T.J.'s idea, not his."

"No, Wanda's up there already and DeeDee's there too. I want them both to be with me when my time comes, and it will be much easier to go through the last four months in the lighter gravity up there." T.J. said, knowing that Wanda and Rudy had taken the apartment next to theirs so Wanda could be close by if she needed her.

"Don't forget that Addi and Zithri are just across the hall." K'Narf said, as he opened the door and helped T.J. into her seat.

"You're not planning to stop at the Whale are you?" T.J. asked, as K'Narf ran around and climbed into the driver's seat.

"No, everything we had there has already been packed up and sent ahead of us. Did you get everything squared away with the banks like you wanted?" K'Narf asked, as he cloaked the Lady and lifted off.

"Yes, I have good relations with several bankers that also happen to be very old and dear friends. They are going to see that the general fund stays well funded just in case I don't return in time to make the necessary transfers myself. I left certain pieces of collateral in their collective possession so they would not feel that I was taking advantage of our friendship. Not that I don't trust them to make the transfers on my behalf, but I've never trusted my money to anyone and I don't intend to start now." T.J. said, thinking of the priceless jewelry she had left in their collective vaults as potential collateral.

"How much are you leaving in the general fund?" K'Narf asked, as he started increasing thrust once they cleared the satellite belt.

"I have seven hundred million in it now, and they agreed not let it go below three hundred million, so whatever our needs the money will be there until I can get back."

"That's a lot of money." K'Narf said, thinking of how hard it had been for Rudy and him to scrap up a few hundred thousand hustling the casinos.

"Yes, it is. The Lord gifted me with the ability to greatly enhance that which I received at birth. Most of my life I did not understand why, now I do." T.J. said, knowing that without her vast resources the building of Etaguf city and the fighter fleet would not have been possible.

"He does have his way, doesn't He?" K'Narf said, referring to God.

"That He does, and so do I." T.J. said, thinking of the many other arrangements she had to make in order to take this extended pregnancy leave. "I have also arranged for Premo and Sylvia to keep their eye on certain stocks for me as well. I really do not foresee any serious problems." She added, trying to get comfortable as the pressure increased to one-point-five G's of thrust.

"Are you sure you can handle this amount of thrust?" K'Narf asked, knowing that Zithri and Addi had said it was okay, but wanting to make sure T.J. agreed now that she was feeling it.

"Yes, if the baby can stand it, I can stand it." T.J. said, tilting her chair back into a more comfortable position. "How long does the computer say it will take us at this level of thrust?"

"Well, we left the minute Sabbath was over, so we should get there Friday evening just as Sabbath begins, of course we picked up twelve extra hours due to the time differences." K'narf said, double-checking the computer settings as he too settled back. It was going to be a long six and a half days, he knew, and he wasn't carrying a baby in his stomach.

"Are you sure you're all right?" K'Narf asked for the hundredth time. Looking over at T.J.'s face. A face that looked strained to him.

What possessed me to let him get me pregnant, T.J. thought for the hundredth time, once for each time K'Narf had asked her if she was all right. "If I say yes, will you believe me?" T.J. asked, turning her head to look at K'Narf.

"I'm just worried, you look so stressed out." K'Narf said, looking at the anger in T.J.'s eyes and knowing he was the cause, but not knowing what else to do except ask her if she was all right.

"Maybe that's because I'm carrying your baby." T.J. said, trying to fight the irritation that kept bubbling up inside of her. She knew it wasn't K'Narf's fault; he was just the only one she could blame unless she wanted to blame herself and that was ridiculous. "How much longer?" She asked, thinking they ought to be getting close.

"We're coming up on Pluto now." K'Narf said, pointing out through the front windshield.

"Thank God, I don't know how much more of this I could stand." T.J. said, struggling to sit up enough to raise her seat into the upright position. "It's a good thing that they have that treatment plant up and operational because I need a shower, a very hot shower." She added, feeling K'Narf ease up on the reverse thrust as they approached the planet.

"They're waiting for you." K'Narf said, having received a flash from Wanda. "I'm supposed to set the Lady down right at the door to the apartment building. Wanda, DeeDee and Redwing are all there waiting. They even have a wheelchair ready for you." He added, smiling at the look on T.J.'s face at the mention of a wheelchair.

"Are you all right?" Wanda asked, as K'Narf opened the Lady's door so everyone could see T.J.

"If I hear those words one more time I'm going to scream." T.J. said, struggling to turn around so she could get out of her seat as Wanda reached for her.

"Yes ma'am." Wanda said, glancing at K'Narf as she helped T.J. out and onto her feet.

"And I don't need a wheelchair!" T.J. exclaimed, grabbing Wanda's arm and then Redwing's as they headed for the open door hallway that led to the apartment. "What I need is a good hot shower. I know that this suit keeps me microscopically clean, but my mind tells me I still need a hot shower and lots of soap." She added as the three of them, with DeeDee bringing up the rear with the empty wheelchair disappeared into the building's interior.

"That must have been one rough trip." Karl said, looking at the strained look on K'Narf's face.

"You have no idea." K'Narf said, turning to look at Karl. "Climb in, let's get the Lady back up to the parking area." He added, as he started to head for the drivers side.

"No, just leave her here, it'll make it easier for my guys to get her unloaded. Right now I think it best we leave the ladies alone. Why don't we head for the commissary and get a cup of coffee.

"Sounds good." K'Narf said, as they headed for the stairs. "Where's everybody at anyway?"

"Young Mister Neli is with Brent, and..."

"And where's Brent?" K'Narf asked, cutting Karl off.

"Don't know for sure, but wherever he is Neli's with him. Those two seemed to have formed a bond of some sort."

"Like brothers?" K'Narf asked, glancing at Karl.

"No, more like mischievous friends, anyway it's obvious they like each other."

"And Addi, where's he?"

"Same thing, only he's running around with Rudy. Seems they both like gardening and they both like cooking so you'll either find them in one of the green zones, over at the greenhouse in Brent's ship or in the kitchen bothering the cooks." Karl said, thinking of last week when the cooks got together and threw them both out of the kitchen because of some kind of mess they had made.

"Yeah, Rudy cooked for the Navy for some thirty years." K'Narf said, remembering the stories Rudy had told him about the Navy. "He does seem to enjoy it."

"Yes, they both do, hey, look." Karl said, stopping and turning around. "From here you can see the skeletons of the new apartments that we're putting up." He added, as he pointed at the network of steel girders rising up around the apartment building they had just left. "My guys would be crawling all over those if it wasn't the Sabbath." He added, pointing at the area he was talking about.

"How many are you building?" K'Narf asked,

thinking the area he was pointing out was a whole lot bigger than one building.

"We're doing that whole section where your apartment building is located, now that the design has been proven out, three buildings deep and seven buildings in width. Don't want to build just one big building in case of a serious meteor hit, better chance for survival if the buildings are separate units." Karl explained, turning and continuing up the stairs to the fifth level.

"How many apartments will that be?"

"Well, that first level allows us to put up a building three hundred feet tall, and each of the first level sections, and there are eight of those, can accommodate twenty-one buildings." Karl said, mentally seeing what he was describing as they walked. "So the building's in the section we are working on will have fifteen levels and each building level is laid out with seven various sized apartments so that adds up to two thousand two hundred and five apartments. That will get most of our people out of their tents right now, but of course there are more newbies arriving every day." He added, frowning at the thought of all the newbies coming in.

"How do they get here?" K'Narf asked, having been meaning to ask that question before.

"If they aren't shanghaied by Mitch Murphy or Minh Le, they hitch a ride up with one of my truckers.

"So that's how they're getting up here." K'Narf said, thinking he should have thought of that without having to ask.

"Yeah, my driver's love it, especially when they're girls." Karl added, remembering how one of his driver's had allowed four of them into his cab all at the same time. Of course three days under three point two G's of thrust with four ladies turned out to be a little more than he could handle. At least he never did it again, Karl remembered, smiling at the memory of what little Tommy had looked like when they opened the door to his cab after he landed.

"Be my guest." K'Narf said, opening the door to Redwing's commissary.

"Thank you sir." Karl said, going through the door and heading for the line. "Oh shucks." He added, frowning.

"What's the matter?" K'narf asked, seeing the disappointment on Karl's face.

"Just remembered, today's the Sabbath and the chance of there being a piece of Maude's pie left is as good as the proverbial snowball in you know where." Karl said, as they moved over and got into line.

"What do they serve on the Sabbath?" K'Narf asked, noticing how fast the line was moving.

"Lot of fruit, fried lamb chops and chicken legs, little bags of chips and sweet rolls. Stuff like that, and when it's all gone it's all gone. Nobody here to refill anything." Karl said, as they got close enough for him to see if there was any pie left. "Thank you Lord." He added, as he spotted a piece of pie still sitting on it's bed of ice, then watched as the woman two places ahead of him took it. I am not that fat, he thought, thinking the Lord had taken it away from him on purpose.

"What's wrong?" K'Narf asked, knowing something was bothering Karl.

"Nothing, just not that fond of cinnamon rolls." Karl said, as he picked one up and put it on his tray.

"At least there are plenty of seats now." K'Narf said, as they sat down at an empty table.

"Did anyone tell you about our lending library?" Karl asked, changing the subject as he sipped at his coffee.

"No, what's that?"

"After I loaned you that DVD by Mark Biltz we got to talking and decided we needed a lending library where we could come and check out something to watch when we're on downtime."

"Like what?" K'narf asked, talking around a mouthful of a cold lamb chop, a rare treat since T.J. wouldn't allow any meat in the house.

"That turned into a problem because of the messed up rating system and way too many directors use God's name as a curse word, what we call Satan's Mark, in their movies, not counting a lot of so called Christian preachers teaching things that don't line up with the word of God." Karl said, frowning.

"I think I know what you're talking about." K'Narf said, remembering the movie he had tried to watch that night he'd spent in a New York hotel room where an actor had done just that, taking God's name in vain and forcing him to change the channel. "How did you get around it?"

"We set up a screened and unscreened section. We had our people on the ground go out and buy three of every movie they could get their hands on that they thought might be fun to watch and the same thing for all the preacher's that sell copies of their sermons and anything else that struck their fancy and we dumped it all into a storage area and labeled it '*unscreened*'. Karl said, obviously proud of what they had come up with.

"And then?" K'Narf asked, leaning back in his chair holding his almost empty coffee cup.

"Simple, if someone takes a bundle of three out of the unscreened section and finds that a director has violated the third commandment and used God's name as a curse word or his work represents an immoral story line, they're thrown into the recycle bin. Same goes for anything else if it's deemed unfit to be seen for whatever reason; including a preacher teaching something that doesn't line up with God's word."

"How do you recycle them?" K'Narf asked, thinking of plastic and glass.

"Just to make sure they don't survive somehow, we recycle them into the sun. Sent a batch of movies that way yesterday. Anyway, if you want to play it safe, go to the screened section and get something out of it. That way you won't find yourself halfway through a good movie and realize that you need to stop it and take it to the trash bin, sorry, I mean the recycle bin." Karl said, grinning at his play on words.

"Not to change the subject, but what's on schedule for tomorrow?" K'Narf asked, thinking he'd like to meet with Brent and get himself updated.

"Brent wants all of us over at his control room to go over his battle plans. He's been waiting for you to get here before getting us all together." Karl said, as he stood up.

"Where you going?" K'Narf asked, seeing that Karl was anxious to get going.

"Well, the girls ought to have T.J. settled in by now so it should be safe for you to go home, and I, uh, have a date with my wife so to speak." Karl said, stammering a little as he mentioned his date.

"I take it DeeDee is no longer at my apartment." K'Narf said, pushing his chair back and standing up.

"No, she uh, just flashed me wondering where I was. We, uh, have a little group study of the bible on the Sabbath." Karl said, stammering again.

"How little?" K'Narf asked, beginning to get the picture.

"Where two or more are gathered together." Karl said, a big smile on his face as he walked away, leaving K'narf standing outside Redwing's commissary all by himself. It was the first time that he had ever been left alone in this city that carried his name. A name that he had made up to spite his father, he remembered, as he walked down toward level one and T.J. Marveling as he walked at how much had happened and how much he had missed in just the few days he had lost bringing T.J. up here. He also found himself looking forward to the meeting with Brent tomorrow as he looked at the city that even with all it's lights on, still seemed to be dimly lit.

Chapter Twenty

"Ah, your here." Brent said, as K'Narf walked into the control room, which was now all fired up with pulsating color and movement on almost every screen. "As you can see all of our positions are now manned, or peopled I should say since most of them are ladies." He added, waving his arm in a broad sweep that took in all of the individual single screen positions.

"How many fighters do you have?" K'Narf asked, looking at all the fired up screens.

"Enough for every operator to have several fighters assigned to them which allows us to run practice drills with everyone involved. Good, here's Neli and Toby." Brent said, motioning them to come down to the dais area.

"How was your trip?" Toby asked, as he walked over and sat down in the big swivel chair Brent had assigned to him.

"Don't ask if you don't really want to know." K'Narf said, setting down in the "*it*" chair next to Toby.

"Question withdrawn." Toby answered, knowing he really didn't want to know what spending six and a half days in transit with a pregnant T.J. was like. "Where are Addi and Zithri?" He asked, as he spun his chair around and looked at the empty chairs behind him.

"Addi is down in the galley with Rudy cooking something, but he knows we're waiting on him, and Zithri is with T.J. and won't be coming. At least not for today's drill." Brent said, motioning Neli to his seat.

"What exactly have you planned for us today?" K'Narf asked, looking at thirty two large flat screens in front of him, all but two of them filled with what looked like camera shots from the undercarriage of a fighter that was dodging and darting all over space. "How come two of the screens are blank?"

"Those belong to the scout ship launchers and they aren't participating in today's drill."

"How often do you have drills like this?"

"We do this every day. I want all the pilots and support personnel as familiar as possible with their jobs and equipment. You're here today to critique us, that is to see if you can spot anything we're doing that we might improve on, as well as to familiarize yourselves with what's going to be happening in this room when the battle actually starts."

"Why aren't the fighter's cloaked?" K'Narf asked, watching as fighters flitted from time to time across one of the screens.

"They will be when the actual drill starts, we don't cloak until then for safety reasons." Brent said, seeing Addi come into the room and waving him down to the dais and pointing to his chair.

"I thought you said their weapons could not fire on each other?" K'Narf asked, remembering the tagging devices Brent had told him about.

"They can't, but unfortunately they can physically run into each other."

"Anybody hurt?" K'Narf asked, knowing that Brent wasn't talking about a hypothetical situation.

"Only their pride. Did require some rather serious repairs to their Teardrops though." Brent said, remembering just how sharp K'Narf could be when he put his mind to it. "Good, we're all here so let's get this show on the road." He added, as Addi stepped onto the dais.

"This is quite impressive, I take it this was your idea?" Addi asked, as two fighters seemed to be playing chicken with each other on one of the screens.

"I figured if President Obama was smart enough to mount cameras on Navy Seals out on a combat mission and watch them from his situation room in the White House; why couldn't we watch though cameras mounted on our fighters here." Brent said, explaining how he got the idea.

"Ingenious!" Addi exclaimed, taking his seat.

"Thank you, now, what I am going to do is walk you through a graphic mock up of how we think the battle will go down." Brent said, as twelve large black battle ships appeared on the center screen causing Addi to give out a gasp. "Sorry Addi, should have warned you. Here's how we plan to deploy our fighters." He added, continuing on as he backed away from the simulated fleet so you could see the proposed positions of the Teardrop fighters.

Our battle plan is rather simple, when they come into range our two scout launchers will fire their paintball missiles, six at each warship, which will look like this." Brent said, showing six missiles suddenly cork screwing in at each of the big black battle ships. The simulation showed the battle ships firing with only one missile getting through out of each barrage of six missiles.

"We will still have twelve additional missiles available if these don't do the job, only they will require our pilots to get close to the warships and fire them at close range which we do not want. Any questions so far?" He asked, looking at each of them in turn.

"Will one paintball missile do the job?" Toby asked, wondering how one missile could do that much damage to a ship that large.

"Yes, one missile hit will create a huge cloud of paint particles that will envelop the entire ship and the magnetic particles embedded in the paint will draw the paint down and literally coat the ship blinding any optics they may be using and jam, or at least create static, on any telecommunication equipment they may have as well."

"Thank you." Toby said, thinking his question had been answered, thinking also that Brent was going for overkill with that many paintball missiles.

"This first attack should draw out their fighters which will be taunted by three of our Teardrop fighters. What we want to do is draw the fighters away from the protective cover of the warships just in case they can still see and fire. This will also allow our two scout launchers to move in and drop the black widows onto the hulls of the warships."

"How close do they have to get?" K'Narf asked, as the simulation showed fighters pouring out of the battle ships and chasing three uncloaked Teardrop fighters.

"Unfortunately they have to come within a mile of them before they can release the widows." Brent said, noticing that it was K'Narf that had spotted the only weak point he was aware of in his attack plan. "After their fighters are a sufficient distance away from their mother ships the three fighter decoys will cloak and we will attack them with everything we have. We are praying that we can destroy over half of them before they even realize who or what they are fighting. Questions?" He asked, turning around in his chair so he could look specifically at Addi.

"I have one." Neli said, timidly raising his hand.

"Please." Brent said, wondering what Neli could have spotted. Knowing his knowledge of the Snark fleet to be even less than his own.

"Uncle Addi, you once told me that those ships that destroyed my parents city had flying machines that looked like Brent's scout ship launchers, and that they had scoured the planet looking for us after they threw that asteroid at the planet. Remember, that's when you and uncle Zithri flew your ship into that cave so they couldn't see us." Neli said, dragging up childhood memories of sitting on his uncle Zithri's lap when they were hiding out. "Aren't they considered fighters too?"

"Addi." Brent said, looking at Addi. "Do those warships have five man scout ships as well as two man fighters?

"Yes, I'm afraid they do."

"How many?"

"Each warship should have two of them, and in answer to Neli's question, yes they are considered to be very deadly fighters when not being used as a troop transport.

"That puts a new spin on our battle plans." Brent said, looking worried.

"Why, what do you mean?" Toby asked, knowing something bad had just gone down.

"We now have twenty four heavily armored five man fighters that we are going to have to deal with along with their nine hundred and sixty two man fighters." Brent said, feeling slightly over whelmed. "And there is no way at this late date to increase our fighter production. We are barely going to make our three hundred production mark as it is."

"You say heavily armored, will your missiles penetrate their armor?" K'Narf asked, thinking this to be a critical consideration.

"Not as they are programmed now." Brent said, still frowning.

"What do you mean programmed?" Toby asked, thinking of artillery shells rather than missiles.

"All of our missiles are currently set to explode on contact due to the thin skin of a two man fighter. If we fired them at one of those scout ships they would explode on contact and be about as effective as throwing firecrackers at them. What I mean is that they wouldn't penetrate their armor before exploding." Brent said, trying to explain the problem.

"Can you fix them?" K'Narf asked, again getting at the core of the problem.

"Yes, but it will require a jury-rig to the firing console in all of our fighters and we don't have a lot of time. What I mean is that we will have to redesign all of our missiles so they can be programmed on the fly by the gunner." Brent said, already seeing the redesign in his mind.

"May I make a suggestion?" Addi asked, instantly getting everyone's attention.

"Please." Brent said, giving Addi the floor.

"In the battle you have described, the two man fighters will come out first simply because they can be manned quicker. This means that they will already be miles away chasing your decoys by the time the scout ship fighters can be manned and launched. If you were to have some of your fighters cloaked and waiting for them you may be able to knock them out with a surprise attack just as they emerge from their warships. Would this be possible?" Addi asked, looking unsure of himself.

"Yes!" Brent exclaimed, instantly seeing the new battle plan in his head. "And after they take out the scout ships they could quickly reprogram their missiles and join the battle by attacking the Snark two man fighters from the rear."

"Do you have enough time to make the modifications?" K'Narf asked, thinking of the already overworked assembly lines.

"Yes, I'll get my people on it immediately, we can do the modifications right here so it won't interrupt Karl and Mitch's people." Brent said, seeing the relief in K'Narf's eyes as he said they wouldn't need to bother the assembly lines.

"Could I make one other small suggestion?" Addi asked, drawing everyone's attention back to himself again.

"Believe me, you have our full attention." Brent said, giving Addi a nod of respect.

"You told me about that Carthy, I'm sorry, Snark officer that committed suicide when he saw

Mister Etaguf with his eyes glowing like one of the space people. This leads me to believe that their leaders have filled them with a fear of the space people to such an unreasonable level that they might panic at the thought of actually being in a fight with them."

"What are you trying to say?" Toby asked, thinking he knew but he wanted to hear Addi say it in another way.

"What I mean is that if you can make them believe they are being attacked by space people, they will panic and a fear driven man does not make a good fighter."

"And how do you suggest we do that?" Brent asked, liking the idea, but not seeing a way to do it. The Snark officers already knew who they would be coming up against from their probe reports, and it wasn't the space people.

"Your Teardrop fighters, when not cloaked, look very similar to the small two man ships the space people use. They are both shiny and silver looking and if a squadron of at least six uncloaked fighters attacked each of their warships with a barrage of missiles into the hanger bay immediately after your paintball missiles' hit, and allowed the emerging fighters to see them before they took off, they just might think they were being attacked by the space people. On top of that your barrage of missiles into their hanger bay may cause serious enough damage to cut down the odds a little." Addi said, taking a deep breath.

"It was my understanding that neither you nor your brother had any military training. Did I understand wrong?" K'Narf asked, putting into words what both Brent and Toby were thinking.

"I have lived a long time and have seen many wars among your people, I may have picked up a pointer or two during that time." Addi said, pleased that his ideas were being considered.

"Let me put it this way, if you have any more ideas I want to hear them." Brent said, his mind instantly adding this new facet to his battle plan.

"Is this possible?" K'Narf asked, wanting to know if Brent agreed.

"Yes, it will mean training two new attack formations as well as learning to reprogram missiles on the fly, but it can be done."

"Do you have enough fighters?"

"Yes, I think I do, I will need seventy two for the attack on the hanger bays, but they will leave immediately and lead the Snark fighters into the trap being set by the rest of our fighters, and I will need forty eight fighters to stay cloaked and wait for their scout ships to come out. Then, once they take out the Snark scout fighters, they can reprogram their missiles and take on any Snark fighters trying to get back to their mother ships as they join up with our other fighters." Brent said, the new battle plan firming up in his mind.

"When can you show us this new battle plan?" K'Narf asked, instinctively knowing that this meeting was over.

"In thirty days. I'll have everything done and ready for another show and tell in thirty days." Brent said, repeating himself. "And I want your brother here next time too, tell him no excuses." He added, making Addi smile the biggest smile K'Narf had seen on his face since they had first met.

"I take that as a compliment, and yes I will tell him." Addi said, still smiling.

"And Neli, thank you for remembering about those Snark scout ships. You may have just saved all of our lives." K'Narf said, noting that it was Neli's memory that had triggered this new battle plan as Neli's face lit up with a smile that mirrored his uncle's.

"Yes, yes it may just have." Brent said, agreeing with K'Narf as they all stood up. "Oh, and by the way, I have a message for you from your father." He added, taking K'Narf off to the side so they could have a moment of privacy.

"What?"

"It seems that your mother is madder than a wet hen because you didn't invite her to her first grandchild's birth, and haven't even had the decency to bring the baby by for her to see it." Brent said, trying not to smile as he delivered the message.

"But there isn't a baby yet."

"She's counted the days and as far as she is concerned you're toast." Brent said, again trying to conceal a smile as he elaborated on K'Narf's message from his father.

"Uh, Mister Etaguf, Mister Conrad." Addi said, calling over to them to get their attention. "Would you two like to join us in the galley. Rudy and I have put together a quiche using one of the new vegetables from our world and it should be coming out of the oven just about now. I think you'll really like it."

"Wouldn't miss it for the world." K'Narf said, as he and Brent joined them. "Do you think there might be enough for me to take some to T.J.?"

Chapter Twenty-One

"Are you all right?" K'Narf asked, wondering if it was safe to ask that question again.

"Yes, as a matter of fact I am. I had a long hot shower with plenty of soap, a good long sleep and Wanda came over and we fixed ourselves a very nice and very late breakfast or early lunch or whatever you want to call it." T.J. said, struggling as she got off the couch and gave K'Narf as good a hug as her belly would allow. "I'm sorry I was so rough on you, but that diaper space suit coupled with a very long uncomfortable and sleepless trip was a little too much to for me to handle, forgive me?"

"Nothing to forgive, I just never want to do it again, okay?" K'Narf said, hoping he wasn't being too honest.

"Deal, now how did your meeting with Brent go?" T.J. asked, letting go of K'Narf and setting back down on the couch with his help.

"Good for us, bad for Brent." K'Narf said, sitting down next to T.J. and taking her hand in his.

"What do you mean?"

"Brent didn't know that each of those warships carried two five man scout ship fighters so he's going to have to revamp his entire attack force and some of their missiles."

"Why the missiles?"

"Some of them need to be armor piercing, but that's not what I need to talk to you about."

"Do I want to hear this?" T.J. asked, seeing as well as feeling the uneasiness in K'Narf's face.

"It seems my mother's throwing a fit. My father said she thinks you've already had the baby and that not only wouldn't I let her be there for the birth of her first grandchild, I don't even have the common decency to bring you and the baby over so she can see it. That's about it in a nutshell as my grandmother used to say." K'Narf said, carefully watching T.J.'s reaction.

"There is no way you're going to convince her that I haven't had this baby unless you bring her up here." T.J. said, smiling at the affect her words were having on K'Narf. "What's that?" She added, pointing at the foil covered bowl K'Narf had set down on the coffee table.

"That's a double helping of a quiche that Rudy and Addi made over in Brent's galley. It's made from one of those new vegetables Rudy's growing over there. It'll have to be heated up a little, but I think you'll like it I know I did." K'Narf said, not really wanting to change the subject. "Bring her up here, there must be another way."

"I'm not due until the middle of August and that's a little over two months away. There's no way that we can put her off that long." T.J. said, looking at K'Narf, knowing that he knew she was right.

"Would it be safe, I mean we might be attacked." K'Narf said, struggling for a rational objection to a visit by his mother.

"If they get past Brent's fighters it would be

more likely for them to grab the Whale and throw it at the Earth than attack us here." T.J. said, destroying what little logic there was in K'Narf's objection. "And she already knows I'm not a run-of-the-mill daughter-in-law. So if you were careful in what you told her, she shouldn't be too shocked when you lift off since she's seen you do it once already."

"Yeah, but she didn't know we flew into space, and I'd have to convince her to pack for an extended stay. She'd also have to wear a diaper space suit as you call it, which I'd have to get her measured for and..."

"And what?" T.J. asked, breaking in on K'Narf's string of objections. "When you get through telling me why it can't be done, we need to start talking about how we're going to do it."

"Go on." K'Narf said, knowing his defeat was imminent.

"We have a perfectly good guest bedroom she can stay in, and it might be fun having her help me decorate the other bedroom as a baby's room. I also flashed Irena a little while ago and she's upset that Toby spends so much time up here and now that Wanda and I are here she wants to come up too. It might make it easier on you if she came up with you, then your mother would have someone to commiserate with instead of accusing you every time something bothered her or she was uncomfortable like I did." T.J. said, seeing defeat in K'Narf's eyes as she spoke.

"You really want her here?" K'Narf asked, knowing it was a feeble question.

"Let me flash Irena and see if she would want to ride up with you and your mother. Just give me a second." T.J. said, closing her eyes.

"Well?" K'Narf asked, as T.J. opened her eyes.

"She said yes, but she has to train a replacement before she can leave the school. I told her that was okay because you had to get your mother measured for a diaper suit and have it grown and that you would contact her with a departure time as soon as you know it. Meanwhile she is going to do some shopping for me for boy baby things. I'll either save my pink things or give them to Sylvia if it turns out she's going to have a girl" T.J. said, without inhaling even once.

"Pink things?"

"Yes, remember, I thought I was going to have a baby girl. Anyway, now that we know for sure that Sylvia's pregnant, she might need them if it turns out to be a girl." T.J. said, repeating herself for K'Narf's benefit.

"Okay, Brent said that if I compressed myself a little, enough to withstand about six or seven G's, just enough to give me a tan look, he would get one of his pilots to drop me off at Sutton Wells. I can have the Lady tied down on one of Karl's empty flatbeds on its way back to Sutton Wells so it will be there waiting for me. That will cut down on the time I have to be away from you." K'Narf said, letting T.J. know that he had already talked over his mother's coming up at least with Brent and Karl and probably with Toby as well.

"I see, is there anything else that you fellows have talked about that I should know about, feeling through K'Narf's emotions that there was."

"Yes, well, it seems that there is some grumbling among some of the newbies about work assignments." K'Narf said, seriously doubting if he should dump something like this on T.J. in her condition.

"Oh, is that all." T.J. said, looking relieved.

"What do you mean is that all?" K'Narf asked, confused by T.J. reaction.

"Just that we've been working on that for weeks, we're going to open a bank."

"A bank, how's that going to help?"

"First let me say that everyone deserves certain basic things without begging for them, like a roof over their heads and food, things that allow them to live with a certain amount of dignity. On the other hand, they don't deserve the finer things in life without working for them."

"Go on." K'Narf said, listening intently.

"Up here you have men and women risking their lives up on the high steel and working in the treatment plant where, to be quite blunt, it stinks. We also have people working hard hot sweaty shifts cooking in the commissary kitchens, not to mention three hundred newly bonded couples that are getting ready to risk their lives for us as fighter pilots."

"Yes, that's what we were talking about this morning over coffee." K'Narf said, thinking of what Karl and Brent had told him.

"Well, we've been talking about it a lot longer than that and as of Monday morning we're opening a bank."

"I still don't understand, how's a bank..."

"Let me finish. We're going to get everyone to open an account using his or her fingerprints as their account number. Starting out, and each month after that, every account will be credited with one hundred dollars whether they work or not. In addition to that, each person will be entitled to an apartment depending on their size, efficiencies for singles, a one bedroom for couples and two and three bedrooms for families. Nobody will be charged rent, electric power or water. All these items will be considered basic necessities." T.J. said, taking a deep breath. "Not terribly unlike what the United States tries to do with their subsidized housing and food stamps." She added, exhaling slowly.

"We have a three bedroom." K'Narf said, still thinking of themselves as a couple.

"They will also be entitled." T.J. said, ignoring K'Narf's comment. "To basic food supplies from the commissary's pantry and once the commissary is turned into what amounts to a soup kitchen they will be allowed to eat there without paying."

"But they don't pay now." K'Narf said, confused.

"No they don't, but once everyone has an apartment where they can do their own cooking the menu will be changed to a basic menu similar to what you find today in free soup kitchens in city's like New York where they can eat if they don't feel like cooking."

"I still don't see how that is going to solve the problems Karl and Brent told me about today." K'Narf said, seeing how everyone could and would be treated with respect and dignity, but not how that would help the problem of people who felt that they were being taken advantage of due to their work assignments or risks they were taking.

"Because we are going to set up a committee that will establish an hourly wage for every category of job there is in this place, including Mayor of the

City." T.J. said, smiling at K'Narf. "And they will receive back credit in their account for all the hours they've worked."

"Credit, but what can they spend it on up here?" K'Narf asked, still not seeing what T.J. was driving at.

"You remember that mural Debbie painted over in Redwing's commissary. The one where she painted in all those shops built over all the combination green zones and walkways, and how there is going to be shops for each level in the city, not to mention what's going to be built down in the center of the city. Well, those shops are going to be selling things, like jewelry, clothing, fresh vegetables, spices..."

"Spices?"

"Yes, the only spices that are considered basic are pepper and salt. Everything else will have to be purchased if you want it. The emphasis being on want, versus need." T.J. said, driving home her point.

"Then all I have to worry about is what they're serving at the soup kitchen. Doesn't sound all that bad." K'Narf said, having figured out what T.J. was telling him about how this new system was going to work as he tried not to openly grin.

"Did I mention that many, many of those little shops will be specialty restaurants and that I don't like to cook, or that your new son will be needing a baby sitter when you take me out to eat. Not to mention the monthly fee we're going to be paying to Addi and Zithri, probably for the next eighteen years." T.J. said, smiling openly at K'Narf now that he was beginning to appreciate the intricacies of the new system she was getting ready to put in place.

"Has anybody taken that 'Mayor of the City' job yet?" Thinking that might be a decent paying job.

"No, but it's going to be an elected position so you'll have to run for it." T.J. said, then added, a smile still on her face. "Of course, one of the highest paying jobs will be working in the treatment plant, but you'd have to take a shower every day before coming home. I wouldn't let you into the house smelling like that place."

"I want to see the list of jobs and their pay scale as soon as it's available." K'Narf said, more seriously than he intended, thinking it was possible he might not get that "Mayor of the City" job, even if the city was named after him. "Would that 'Mayor of the City' job be a lifetime position?"

"I don't know, we haven't talked about that yet." T.J. said, realizing that K'Narf was actually serious.

"Do I get to be on this committee?" K'Narf asked, trying not to look at T.J.

"We had planned on asking everyone you've put on the council to be a member, why?"

"Just that you can't influence something if your not allowed to talk before a vote is taken." Knowing now that he really wasn't kidding, as he looked T.J. dead in the eye.

"Somehow I knew this wasn't going to be as easy as it sounded." T.J. said, letting go of K'Narf's hand as she said it, knowing a politically tainted statement when she heard it. "You need to get a message to your father and tell him when you'll be coming by to see them. How do you do that anyway?" She added, asking the question as she changed the subject before she found herself talking in Mandarin.

"Through Minh Le, why?"

"How long did Brent say it would take his pilot to get you to Sutton Wells if you compressed yourself?"

"About two days, why?"

"If you compressed yourself the rest of this week, then you could be ready to go by the time the coming Sabbath is over, couldn't you?"

"Yes." K'Narf said, starting to get upset because of T.J.'s refusal to answer his questions.

"Good, then you need to flash Minh Le and tell him to tell your father and mother that you'll drop by to see them Tuesday evening sometime after sundown, say around nine thirty or ten o'clock."

"You're absolutely sure you want me to do this?" K'Narf asked, demanding an answer to his question as he sat up and turned to face T.J., thinking how small and helpless she looked, then immediately thinking how deceiving looks could be.

"Yes, I believe there is no other way." T.J. said, looking at and feeling the uneasiness in K'Narf's eyes." And I think we might both enjoy her visit as well." She added, smiling at her obviously unconvinced husband.

Chapter Twenty-Two

K'Narf hesitated as he looked at the doorbell. It was the doorbell to his mother's apartment and he had spent the last month stalling, procrastinating, dawdling and any other synonym he could think of to postpone this moment but it was finally here. Sabbath had ended just hours ago and he had picked up Irena with all her trunks and stuff and flown strait here. He had put four seats in the Lady and filled the passenger seat up front with some of Irena's stuff so his mother would have to sit in the back with Irena. That way they could talk and complain to each other all they wanted without involving him, at least that was the plan. He knew he was going to have to push the Lady up to at least one point six G's just to get there before the next Sabbath started, and that meant a six and a half day trip any way he wanted to slice it, he thought, as his hand involuntarily reached out and pushed the doorbell.

"It's about time you got here." Robbie said, opening the door.

"Hello mother." K'Narf said, as he stepped in and closed the door. "Why aren't you wearing your suit?" He asked, opening his coat letting his mother see that he was wearing his, shiny collar and all.

"I figured I could put it on when we get to wherever it is your taking me." Robbie said, glaring at K'Narf because he had steadfastly refused to tell her where they were going. All she knew was that she was supposed to be there in time for her first grandchild's birth or she wouldn't be going at all. "Now help me with my luggage, I assume that you have that black van of yours parked upstairs again?" She added, asking her question as she picked up one of her smaller suitcases.

"We aren't going anywhere until you put your suit on, I told you that before." K'Narf said, glaring back at his mother and the custom made set of lavender exercise sweats she was wearing.

"I don't like wearing something that fits so tight it doesn't allow me to wear underwear, and I didn't like the attitude of that little man that took my measurements and made it for me either." Robbie said, standing her ground as she presented her arguments.

"His name is Minh Le and he did it as a favor to me, not to you. And for the record where we are going there are no servants, if someone does something for you it's because they like you or because you're my mother. Now go put that suit on, we don't have time to waste arguing!" K'Narf said, raising his voice.

"Mother, do what he says." Joe senior said from the kitchen doorway. "And you don't need to raise your voice at your mother like that." He added, taking a sip of something that definitely was not coffee.

"Yes sir, but we are in a bit of a hurry, we have a long trip ahead of us." K'Narf said, as his mother picked up one of her suitcases and headed for the bedroom.

"So I understand, and it's so secret that you can't even tell your father where it's at?" Joe senior said, asking his question as he took another sip.

"I'm sorry father, mother can tell you all about it when she gets back. Other than that I'd prefer not to say anything at the moment." K'Narf said, thinking if his father knew where he was taking his wife he wouldn't let her go.

"And she'll be gone for at least three months?"

"Yes sir, T.J.'s not due until the middle of next month so I doubt mom will want to come back until Leecal is at least a month old."

"Leecal, that's what you're naming him?" Joe senior asked, looking surprised. I don't think I've ever heard that name before.

"No sir, its a compound of my great grandfather's middle name and T.J.'s father's middle name." K'Narf said, as the door to the bedroom opened. "We like it." He added, as his mother stepped out wearing the suit with its silver helmet collar shining in the light from the overhead chandelier and wearing a dark green full length car coat covering the suits more revealing features.

"Did I hear you say you've already chosen a name for my grandson?" Robbie asked, stepping over and putting her suitcase down with the others. A suitcase that now contained a custom made set of lavender exercise sweats. "Put that drink down Joe, you're going to have to help with these suitcases, they're too many for K'Narf to carry by himself." She added, as she picked up her overnight case and one of the small suitcases and stepped over and opened the door, leaving the two of them to follow her out.

"And who is this?" Joe senior asked, as K'Narf opened the Lady's side door and his father caught sight of Irena.

"Just 'cause she's pretty doesn't mean you have to make a fool of yourself. Now help K'Narf get these suitcases loaded." Robbie said, having seen the affect Irena was having on her husband. K'Narf had told her there was going to be someone else going with them, but he had failed to mention just how gorgeous she was going to be. "And you had better behave while I'm gone. I've told Ada Louise to keep her eye on you and she'll tell me everything when I get back and don't you forget it." She added, as she climbed in and took her seat next to Irena.

"I'll have her back as quickly as I can father." K'Narf said, closing the door and opening his own driver's door.

"No hurry son, no hurry at all." Joe senior said, as K'Narf closed his door and pushed the Lady's cloaking button and lifted off, leaving his father looking just as dumbfounded as he had been that day out at the farm.

"Where are we?" Robbie asked, her voice coming out soft and almost squeaky.

"We're in space mother." K'Narf said, reaching over and programming the onboard computer for Pluto and Etaguf City.

"It's all right Mrs. Etaguf, there's nothing to be afraid of." Irena said, reaching over and taking Robbie's hand.

"Mrs. Etaguf, who's that, oh, you mean me. I uh, K'Narf where are you taking me?" Robbie asked, her voice getting a little stronger.

"To the planet Pluto mother, that's where

T.J.'s at." K'Narf said, enjoying himself even though he knew he would have to apologize and ask forgiveness the next time he talked with the Lord.

"Pluto?"

"Yes mother, Pluto, and it's going to take us a little over six days to get there, and you're going to start feeling a little heavier as we get going." K'Narf said, as the computer kicked in Lady's drive cores and their weights increased up to the equivalent of one point six gravities.

"It's all right Mrs. Etaguf, it's really not all that bad once you get used to it." Irena said, still holding Robbie's hand."

"Don't call me that, my last name's Fugate not Etaguf and what do you mean it's going to take us six days to get there?"

"You heard me mother." K'Narf said, glad that his mother was sitting directly behind him so he couldn't see her face as he looked at the computer readout. "It'll take us exactly one hundred and fifty one hours and thirty seven minutes to be exact, so you might as well relax and get some sleep if you can." He added, trying to get into a more comfortable position himself.

I think I need to go to the bathroom, where is it?" Robbie asked, feeling the increased pressure on her bladder as she looked behind her at the fully packed rear of the van.

"That's her, that's Pluto we're coming up on now." K'Narf said, feeling his mother crane her neck around the back of his seat so she could see out through the windshield.

"Shouldn't a city have lights?" Robbie asked, her voice sounding meek and worn out.

"It does mother, but it's under ground."

K'Narf said, as he took over the controls and dove the Lady for the area where he knew the dome was located. "They've cleared us to come in." He added, as he spotted the dome and dove for the gully that led to the force field entrance.

"Is Toby going to be there?" Irena asked, sounding a little worn out herself. Having spent the last six days holding Robbie's hand.

"He'd better be." K'Narf said, thinking of all the trunks and suitcases that Irena had brought with her. "We're going to land right next to the apartment so it'll be easy to unload everything." He added, as they went through the force field and his mother and Irena saw the city for the first time.

It had changed considerably since the first time he had seen it and that hadn't been that long ago, he thought. All twenty-one buildings in the section he and T.J. lived in were now complete and the next section was under construction. The second level steel cover dome had been completed over the center allowing the water spout from the water fountain down below to come up through it, but nothing was built on it yet. There were also a lot more lights now, he thought, as he maneuvered the Lady to a landing just outside the entrance to their apartment building where he had landed her last time.

"Is that for me dear?" Robbie asked, as K'Narf opened the side door and she saw DeeDee standing there with the wheelchair.

"Yes, I thought you might need it after such a long trip." DeeDee said, flipping up the foot rests so it would be ready for Robbie to set down as K'Narf helped her out.

"Oh thank you, thank you. Do you people

have showers or maybe a bath tub up here?" Robbie asked, collapsing into the wheelchair as K'Narf let go of her.

"Yes, of course we do, and we have a very nice guest room fixed up for you as well." DeeDee said, lifting Robbie's feet one at a time as she put down the foot rests then turned her around and wheeled her into the hallway.

"Rough trip?" Toby asked, as he reached into the van and helped Irena out, giving her a serious hug as he did it.

"You owe me more than one!" Irena exclaimed, hitting K'Narf as hard as she could on the arm.

"I don't know what you're talking about." K'Narf said, grinning as he backed out of harm's reach.

"Hit him for me!" Irena said, turning on Toby. "As hard as you can!" She added, as she took another swing at K'Narf while still hanging onto Toby, missing him by inches. "I'll knock that grin off your face, you just wait."

"That bad, uh?" Toby asked, thinking he'd have to go get DeeDee's wheelchair for Irena too.

"He put her in the back with me on purpose. He knew I'd have to be the one to comfort her and listen to her all the way out while he sat up in that front seat of his and slept, so yes it was that bad." Irena said, trying to hang onto Toby as she took another half-hearted swing in K'Narf's direction. "Where's that apartment you told me we have?" She added, pulling on Toby as she staggered in the direction of the open hallway.

"Go ahead, I'll have her stuff stacked outside your apartment door." K'Narf said, motioning at the two guys Karl had sent over to help them unload as he reached in and grabbed one of his mother's suitcases.

Chapter Twenty-Three

"Brent just flashed me." K'Narf said, setting up in bed and stretching. "He wants me to come over to the command ship, says he's ready for another show and tell."

"Then you need to get over there." T.J. said, struggling as she sat up next to him.

"What about mother?"

"She'll be just fine, she took a long hot bath and then a shower before she went to bed. She got up once to eat during the Sabbath and then went back to bed so she should be feeling pretty good this morning." T.J. said, wincing as she felt a rather strong kick in her stomach.

"Getting kind of rough isn't he?" K'Narf asked, seeing the flash of pain in T.J.'s eyes and then through her he felt the pain caused by his unborn son's kick. "Real rough." He added, as he leaned over and kissed her as his right hand gently touched her swollen belly.

"Yes he is, Zithri helps a lot, he seems to be able to sooth him when he gets agitated." T.J. said, trying to describe what it was the Zithri did. "But you need to get going, they'll all be waiting on you if you don't hurry." She added, pushing at K'Narf to get out of bed.

"Okay, okay, I'll go, I just hate to leave you all alone with my mother." K'Narf said, climbing out of bed.

"You just better pray that we don't become best friends." T.J. said, seeing her threat hit home by the way K'Narf turned and looked at her.

"That's a thought to take with me." K'Narf said, reaching for his clothes and getting dressed.

"It's just that I barely remember my mother, and I find myself wondering what it will be like having your mother around." T.J. said, struggling as she threw her legs over the side of the bed. "You go on now and git out of here before she wakes up. I doubt if she's got over you making her wear that diaper suit yet."

"Yeah, I know. I heard about that all the way out here." K'Narf said, remembering one of the reasons that Irena wanted to hurt him.

"Oh, and I wouldn't count on Irena's vote when you run for Mayor." T.J. said, making K'Narf turn and give her one of his "I don't believe what you just said" stares just before he closed the door and left. He thinks I'm kidding, but I've never heard her that mad in all the years I've known her, T.J. thought, remembering the flash she had gotten from Irena yesterday.

So, now I have to start thinking like a politician, K'Narf thought, as he headed up the stairs toward the landing area where he knew a fighter was waiting for him. That's going to take a lot of fun out of life, worrying about losing someone's vote every time you open your mouth or set someone up like he had done to Irena.

"Gentlemen." He said in what seemed like

only minutes later as he walked down toward the dais and his "it" seat, and like T.J. had said; they were all waiting on him. "Sorry if I kept you waiting." He added, taking his seat.

"Not a problem." Brent said, as he spun around and looked at Zithri. "I understand that Addi has filled you in, so I won't bore everyone by repeating everything that happened last time." He added, as he spun back around and looked at all the lit up screens in front of him.

"Our battle plan is a little more complex than what I originally thought it would be, but overall it is still rather simple. I have created another graphic mockup of what we believe will happen. First the paintball missiles are fired and as soon as the battle ships are engulfed in a cloud of paint particles our attack fighter group led by Dan and Susan Ripplinger will move in and fire a barrage of thirty six missiles into each of the hanger bays and then back off and uncloak."

"Won't that exhaust their missiles?" Toby asked, thinking that was an awful lot of missiles to fire as an initial volley as he watched them do just that in the graphic mockup on the screen in front of them.

"No, we don't think so. Each fighter involved will only be firing six missiles. That still leaves them with eighteen front firing missiles and six that can be fired aft and don't forget they are all armed with twin lasers for close in fighting." Brent said, as alien Snark fighters began to boil out of the warships on the screen. Then, seeing Dan and Susan's regrouped and uncloaked bright silver looking fighters running away they immediately took off after them.

"They will take them to a rendezvous with

the fighters being led by Hacker and Lena where they will fire all of their rear missiles in a concentrated barrage at the pursuing Snark fighters, and then cloak themselves and circle around and join in on the broadside ambush being set up by Hacker.

"Hacker, I thought all of your fighter teams had to be bonded couples?" K'Narf asked, knowing Hacker did not have a bond mate.

"He is, Lena and him bonded four weeks ago. Due to the changes to the battle plan I needed a third team to lead my ambush squadron and when he flashed me that he and Lena had bonded and wanted to join us, I accepted immediately. Hacker is a natural born warrior and I trust him implicitly." Brent said, turning to look at K'Narf as the larger five man scout fighters started coming out of the mock warships on the screen.

Then they all watched in silence, as missiles seemed to come out of empty space and slam into the scout fighters, destroying them all in a matter of seconds. Moments later, in a close-up, three small black objects emerged from empty space and floated toward the nearest warship.

"Those are our black widows. Three of them will be put on each warship where they will seek out and kill anything that moves or has a body temperature above 80 degrees." Brent said, as they watched the small black dots attach themselves and extend eight legs each and start scurrying across the hull.

"Our two scout launchers will join up with Peter and Antoinette Vanlieu's attack group and engage any remaining Snark fighters from the rear, making sure none of them get back to their mother ships. Questions?"

"Yes, how do you shut down those black widows of yours?" Addi asked, thinking something like that running around loose would make it impossible to salvage anything from those destroyed ships."

"First they can be shut down via software. If that fails an implanted explosive device built into each of them can be activated to destroy them. They are dangerous and we have taken appropriate precautions." Brent said, appreciating Addi's question.

"Anyone else?" Brent asked, looking at everyone in turn. "Then K'Narf, I would like for you to come with me. There are some people that you need to meet before this battle of yours actually takes place." He added, standing up.

"Who?" K'Narf asked, standing up as well, knowing that the next time he sat down in Brent's "it" chair it would be for the real thing.

"My pilots, the people that are going to be laying their lives on the line for you. I've arranged for you to meet them."

"All of them?"

"Yes, all of them. They are going to present themselves to you one team at a time and I'm asking you to shake their hands and tell them that you will personally keep them in your prayers. Remember, the odds are that you will never see all of them again unless you go to your chapel down in the Whale and look at their pictures after Debbie gets them painted." Brent said, as he ushered K'Narf toward the exit at the top end of the dais.

"Keep that in mind when you talk to them." Toby heard him say as the two of them disappeared through the exit. "Uh, Mister Zithri, could we talk a minute?" He asked, as Addi and Neli headed out the door with Rudy who had come down to watch.

"Yes of course." Zithri said, stopping and looking at the serious expression on Toby's face. "What is it?"

"Well, It's just that K'Narf asked me a question the other day that has been bothering me, and I thought you might be able to tell me what he meant by it and if he's right." Toby said, giving Zithri a somewhat incoherent explanation of what he wanted.

"What exactly was his question?"

"He asked me why I was still rehearsing for something that had already taken place. I mean I had told him Irena and I were getting ready to clean our house for the feast of Unleavened Bread and that's what he said, I mean asked me." Toby said, still more incoherent than coherent.

"I see, and you want to know if what he asked you has any merit?" Zithri asked, motioning for Toby to walk with him as he headed for the exit.

"Yes, yes, that's it, is he right? He seemed so positive when he said it, I just wondered." Toby said, happy that Zithri seemed to understand what he wanted.

"You and your Irena are both Jewish?"

"Yes, we met while serving in the IDF."

"IDF?"

"I'm sorry, IDF stands for the Israeli Defense Force. We both served."

"I see, and both your parents were Jewish?"

"Yes, we can trace our ancestry back thousands of years, why?"

"Just curious. But before I address Mister

Etaguf's question I would like to ask you a few questions. Is that okay?" Zithri asked, turning to glance at Toby as they walked.

"Yes, of course." Toby said, wondering what kind of questions Zithri could have for him.

"Do you remember reading where God told Moses to declare certain feasts and convocations unto your people and that they were to be considered His feasts and His convocations?"

"Yes, that's in Leviticus twenty-three, why?"

"And your word for feast is Moed is it not?"

"Yes."

"And does not Moed, in the Hebrew tongue actually mean an appointment for a future time, like a set time, such as an appointment to be married?"

"Yes, I suppose it could."

"And does not the word convocation mean rehearsal in the Hebrew tongue?"

"Yes, but it also can mean a public meeting or something called out like an announcement." Toby said, getting confused by Zithri's questions.

"Yes it could, but let's just suppose for a minute that what God said to Moses was that he wanted to set seven divine appointments for a future time with the Jewish people, and that he was going to give specific instructions, convocations, as to how your people were to rehearse for those appointments. Based on your understanding of those two words would you concede that it might be possible that that's what God had in mind?"

"Yes, it could be possible, but I..."

"Two thousand years ago." Zithri said, continuing on as though Toby hadn't tried to answer him. "On the exact day the whole Jewish nation was being told by the priests that they had to clean their houses and get all the leaven out, what did Yeshua do in the temple?"

"You mean when he drove out the money changers and the dove sellers?"

"Did he not clean His father's house on that day?"

"Yeah, I guess so."

"There is no 'guess so' about it, He cleaned His father's house that day, and when He entered the temple through the Eastern gate the next day riding on that colt of His, what Psalms did the people recite to Him?"

"Psalms one-thirteen through eighteen, everyone knows that."

"And what Psalms were the priest reciting, at the exact same time Yeshua was entering through the Eastern gate, to the High Priest as he brought the sacrificial lamb for Israel in through the Northern gate?"

"I see what you mean, they were reciting the same Psalms to the High Priest, weren't they?"

"Yes they were, and the priests were instructed by God to examine that lamb for four days to make sure it was blemish free, how many days did they examine Yeshua?"

"Four days." Toby said, beginning to see what Zithri was driving at.

"And did anyone, the false witnesses, Pilate or Herod find any fault in Him during those four days?"

"No, none."

"And the very hour the scriptures say he was nailed to the cross, was that not the same exact hour that the High Priest tied their unblemished lamb to the alter, and at the exact hour according to the scriptures that Yeshua said 'it is finished' and died, did not the High Priest slit that lambs throat and lay it on the alter and say 'it is finished'?"

"Yes, yes, that would be true, I've just never..."

"And on resurrection morning, the same morning that Yeshua came forth from the grave, was not the High Priest instructed by God to wave a sheaf of barley, the first fruits from the harvest, as an offering of 'first fruits' to Him and was Yeshua not called the 'first fruits' of the resurrection?"

"Yes, yes, I see what you mean. I really do, I just didn't understand. Why has this never been taught? I mean I...."

"God made your people rehearse for fifteen hundred years, my question to you is how could you miss it? And why is a gentile like Mister Etaguf telling you these things, it seems to me that it should be you telling him, not the other way around." Zithri said, stopping as he looked into an open door. "Ah, good we're here." He added, as he looked into the galley at the ceremony getting ready to start.

"Thank you, I..."

"Anyway yes, Yeshua did indeed fulfill the first three appointments His father made for Him, the Holy Spirit fulfilled His on Pentecost and the last three divine appointments will be fulfilled in the very near future. Does that answer your question? If not we can talk again later, right now I think we ought to join the others." Zithri said, stepping into the room and heading for an empty chair at a table down in the front section.

"I haven't been zinged this hard since I flunked spelling in the first grade." Toby muttered to himself, as he followed Zithri into the galley and looked around for an empty seat.

Chapter Twenty-Four

"Oh good, you're back, easy trip?"

"Fast one, came back with one of Brent's pilots at six G's all the way." K'Narf said, flopping down on the couch next to T.J. "I was afraid to flash you. I never know when you might be sleeping."

"Thank you, I don't need what little sleep I do get interrupted." T.J. said, leaning over and receiving K'Narf's kiss.

"How's his highness?" K'Narf asked, looking at T.J.'s swollen belly.

"He dropped and is maneuvering into position to be born according to Zithri." T.J. said, rubbing at her stomach.

"Any labor pains?"

"No, Zithri showed me how to block them. I could be having them right now and not know it."

"Is that safe?"

"Yes, Zithri said he'd know when it's time. Did you get everything done you wanted to?"

"Yes, the last of the fighters have been completed and all the spare parts and tooling are being shipped up here and Premo and Sylvia have moved into the factory. They're turning half of the factory floor into classrooms and using the other half to park Limos. Did you know that they have twenty-four Limo's running eight trips a day up to the Whale now, and that Minh Le has pick-up teams working all over the world."

"I knew that Edna had to have help. She kicked Ralph out and gave his bedroom in the kitchen to two new girls that are helping her now."

"Where did Ralph go?"

"He moved into your old bedroom with Thomas and Judy. They had another litter of kittens and I put in dibs on one of them for us."

"It won't like it here, remember, it won't be able to fly." K'Narf said, remembering how Thomas and Judy had had a fit when he had tried to move them here.

"It will if we bring it here before it gets used to flying. All the other kittens we've given away have adjusted to gravity environments."

"I hope your right, is mom still in bed?" K'Narf asked, realizing that it wasn't yet six o'clock.

"You don't know your mother very well do you?" T.J. asked, looking at K'Narf with a slight sense of surprise.

"Why, where is she?" K'Narf asked, feeling guilty because he had had to go back to Sutton Wells shortly after dropping her off with T.J. and that had been some time ago.

"Out jogging."

"Jogging?"

"With about a hundred new friends of hers. Seems there are a lot of people here that used to like to jog, and just because they don't need to anymore, they still find it fun to run along with your mother who is trying to keep up her strength so she won't be, as she calls it, a marshmallow when she gets back home." T.J. said, looking with amusement at the expression on K'Narf's face.

"But where does she jog?" K'Narf asked, thinking that Etaguf city didn't have any really good places to go jogging.

"Well, she jogged through our hallway last week, but I asked her to please not do that again. She woke up a lot of people that had their curtains closed, like me, and some of them got more than a little upset."

"They got upset?" K'Narf asked, getting a mental picture of hundreds of people jogging by outside their front door at five in the morning. "It must have sounded like a small earthquake."

"I thought we had been hit by a meteor." T.J. said, smiling at the memory, although it hadn't seemed funny at the time.

"How have the two of you been getting along?" K'Narf asked, almost afraid to hear the answer.

"I know you may find this hard to believe, but she is a delight to have around." T.J. said, amused at the expression on K'Narf's face.

"My mother is a delight to have around." K'Narf repeated, not sure if T.J. was putting him on or not. "Is that just your opinion, or do others feel that way too?"

"Let me put it this way, if you want to be elected Mayor of Etaguf City, it will behoove you to actively court your mother's vote." T.J. said, openly smiling at the look on K'Narf's face.

"My mother's vote, how can she have a vote, she doesn't live here?"

"Karl has promised her and your father a two bedroom apartment in the building next door. She's already ordered furniture for it and is planning on coming back up here to spend New Year's Eve with us."

"She going to spend New Year's Eve with us."

"Yes, she wants to spend Christmas with Junior and Cindy and New Years with us and her new grandson."

"And she plans to bring my father up here with her?" K'Narf asked, thinking it was a good thing he was sitting down.

"Yes, and she started a P.A.C. to help get you elected Mayor."

"She started a what?" K'Narf asked, not being familiar with the acronym P.A.C.

"Political Action Committee."

"Why?"

"I happened to mention that you happened to mention that you might like to be Mayor when the time came to elect one." T.J. said, trying to look innocent. "She has Irena running it for her." She added, seeing K'Narf start at the mention of Irena's name.

"No way, the last time I saw Irena she was trying to get Toby to beat me up, and now she's going to help me get elected Mayor, I don't believe it!"

"Well, your mother did point out to Irena that the reason you elect someone is so you can blame everything that goes wrong on them, and that in a place like this there would always be things going wrong." T.J. said, biting her tongue in an effort not to smile. "That may have played a part in Irena's decision." She added, thinking her tongue might be bleeding.

"My mother did all of this?"

"Yes, they've even designed an absentee ballot for everyone that won't be here for the actual election, like her and your father." T.J. said, giving up and openly smiling at K'Narf's expression. "That's why I said that you might need to actively court her vote if you want to get elected."

"Why, who else would she vote for?" K'Narf asked, knowing immediately that he shouldn't have asked the question.

"Well, to have an election you need to have at least two people run for office, so she's persuaded Karl to put his name on the ballot and run against you."

"But I haven't even said I wanted to run yet, how can my name be on a ballot?"

"Oh, she's already collected enough signatures to put you on the ballot. I think she has the papers you need to sign all ready for your signature just to make it legal." T.J. said, trying real hard not to laugh openly at the look on K'Narf's face. "She's trying to talk the election committee into holding the election right after New Year's Eve so her and your father can be here to vote in person."

"Oh she is, is she!" K'Narf exclaimed, standing up and looking down at T.J.'s all too innocent expression. "And how does she think she's going to get back up here, especially if she plans to vote for Karl!" K'Narf asked, starting to get excited. Thinking he really did want that Mayor's job, especially if Karl wanted it.

"Why don't you ask her?" T.J. asked, seeing Robbie jogging toward their door.

"I will." K'Narf said, hearing the door open and turning to face his mother as she entered the room.

"Well, well, the prodigal son returns." Robbie said, opening the door and seeing K'Narf. "It's about time." She added, closing the door.

"T.J. is telling me that you and dad are getting an apartment up here and plan to spend New Year's Eve with us, is that correct?" K'Narf asked, skipping the skillet and jumping right into the fire.

"Yes, what about it?"

"How do you plan to get back and forth, or is that too much for me to ask?" K'Narf asked, knowing he didn't want to spend six days at a crack hauling them back and forth. His mother was bad enough, but with his father along too, he shuddered just thinking about it.

"Karl and the fellows are fixing up an R.V. for us, that's how!" Robbie said, raising her voice to match K'Narf's as she looked over and saw T.J. wink at her. Knowing immediately they had K'Narf on the ropes.

"An R.V.?"

"Yes, an R.V., a recreational vehicle. Karl and the fellows put a treadmill in it so we can keep up our exercises both coming and going, and Jeanie, our driver, is also a physical therapist and will be there to help me with stretching exercises. There's also a large flat screen so we can watch movies and beds for sleeping on instead of sitting up in a chair for six days like you made me do; and there is a bathroom so I don't have to wear that diaper suit all the time!" Robbie exclaimed, ending her tirade on a high note.

"Karl did all this, what's he trying to do, buy your votes?" K'Narf asked, knowing immediately it was the wrong thing to say.

"After that remark, he might just have them!" Robbie said, glaring at K'Narf.

"I'm sorry mom, I didn't really mean that. It's just that I didn't know any of this was going on."

"And you never will know what's going on if you keep taking off all the time. You do know that you have a very sweet wife that is about to have your baby don't you?"

"Yes, I've been keeping in touch. That's why I'm back so I can be here for the birth." K'Narf said, trying to get himself out of trouble.

"I know all about that flashing stuff, it's not much different than having a cell phone, but that's not like being here where you're needed. Now git out of my way, I have to take a shower before the auditions."

"What auditions?"

"We're auditioning for baritones and basses this morning."

"Why?"

"For the choir, they're going to do a medley of camp meeting songs this Sabbath and we don't have enough baritones and bass singers."

"This Sabbath, where?" K'Narf asked, confused again, trying to think of where a choir could set up and sing.

"Down at the new amphitheater, well it's not really an amphitheater. Karl and the fellows built us a place for the choir to stand and put carpet down for better acoustics and they promised to turn off that fountain during the concert. Hard to hear anything over the noise that water makes." Robbie said, stepping over and taking T.J.'s hand and giving her a quizzical look. "You all right, you're looking awful funny."

"No, I'm fine." T.J. said, not knowing if she was telling the truth or not.

"And if you want a seat, you'd better git there early, Karl only hauled up about ten thousand folding chairs. If you're late you'll find yourself sitting in the dirt on the second level." Robbie said, turning her attention back to K'Narf. "Now git out of my way, your gon'na make me late." She added, just as the door opened and Zithri walked in followed by Wanda and DeeDee who was carrying a large stack of necessary sheets and towels.

"It's time." Zithri said, looking at T.J. whose face was turning red, as she could no longer control the contractions that were starting to rack her body.

"K'Narf!"

"What?" K'Narf asked, knowing it had been Brent that had flashed him.

"They're here!"

"Who's here?"

"The Snark fleet! You need to get over here now!"

"What is it K'Narf?" T.J. asked, seeing the look on K'Narf's face as they helped her up off the couch.

"They're here." K'Narf said, knowing T.J. and everyone in the room would know who he was talking about but his mother.

"Wanda." K'Narf flashed, using the common circuit so everyone could hear him. "Put out the word that everyone is to drop what they're doing and start praying. We have six hundred of our people about to put their lives on the line for us and the least we can do is pray for them, and don't let my mother know what's going on."

"Yes sir." Wanda flashed back, ignoring K'Narf's order to drop what she was doing, knowing it did not apply to her or DeeDee and Zithri.

"Then you have to go, there's nothing you can

do here." T.J. said, getting back into the conversation.

"Go! What do you mean go! You're wife is about to have your baby and you're going to go! Not if I break both of your legs first!" Robbie said, looking at K'Narf like she was about ready to do it.

"No mother, he has to go! Trust me, this is not his choice, nor is it mine, but he has to go." T.J. said, looking at her mother-in-law and grimacing as another pain hit. "Wanda will keep you posted until I'm able to flash you myself." T.J. promised, as she shooed at him to leave and headed for the bedroom door with Wanda and Zithri holding her from both sides. DeeDee was already in the bedroom ripping off covers and putting down fresh sheets.

"She said git, so git!" Robbie said, glaring at K'Narf as she turned and followed everyone else into the bedroom.

Chapter Twenty-Five

"Please Lord, may it be Your will that I never have to set in this chair again." K'Narf prayed, as he sat down in Brent's "it" chair. Watching as all the screens started lighting up.

"Karl, Zithri's not going to be able to make it, so you can have his chair." Brent yelled at Karl, who had just sat down in the balcony.

"Thanks, I'll be right down." Karl said, getting up and heading for an exit portal that would take him down to the lower level.

"Addi, Neli, Toby. Looks like we're all here K'Narf." Brent said, setting down. "They told me that T.J.'s in labor."

"Yes, yes she is." K'Narf said, looking at the black fleet that was coming into view as the cloaked scout launchers appeared to be flying backwards as they watched them. "Why are they flying backwards?" K'Narf asked, wondering if his eyes were playing tricks on him.

"We can't do anything until they come to a full stop."

"Why?"

"We can't attack as long as they are still moving in the direction of our solar system, how would we stop them after the battle is over. They weigh millions of tons each and we want to salvage them, not have them crash into Venus or something." Brent said, explaining what he thought was obvious. "They should come to a full stop any second now." He added, looking intently at the screen in front of him.

"What happens if they don't stop?"

"They will, their Commanders don't know what they're getting into so they'll stop and send out a scouting party. Probably a couple of their fighters."

"And if they do, send out fighters I mean?" Toby asked, having followed K'Narf and Brent's conversation.

"They won't get very far." Brent said, concentrating on the screen in front of him. "Have they come to a full stop?" He asked, speaking to someone K'Narf and Toby couldn't see.

"Yes, they are dead in space!" A voice said, coming out of two large speakers hanging above the flat screens in front of them.

"That sounds like Dan Ripplinger." K'Narf said, listening intently.

"Yes, it is." Brent confirmed. "It's your call, we're ready to fire." Brent said, turning to face K'Narf as almost every one of the operators setting at their consoles turned to look at him.

"Then let's get this done, I have a son I want to go see." K'Narf said, having just received a message from Wanda that he was the father of a beautiful baby boy. "Fire!"

All the screens came to life as seventy-two paintball missiles lit up space as they corkscrewed their way toward the massive black warships.

Almost instantly dark green laser beams began reaching out at the missiles with surprised gunners trying vainly to hit them with little success. Within seconds every warship was fully enveloped in a large dark cloud of magnetized paint particles that quickly began to settle on their hulls.

Seconds later twelve barrages of thirty-six missiles each seemed to appear out of empty space as they streaked toward their targeted warship and disappeared into the black hole that they all knew to be the force field for each of the launching bays. The shield prevented them from seeing the explosions taking place within, but it only took minutes before the first of their fighters began to emerge.

"Uncloak, let them see you." Brent said, speaking directly to his fighters through a small microphone he held in his hand, a microphone that K'Narf had not noticed before. "And group up, we want them to think that they're being attacked by a large force." He added, as hundreds of the small black Snark fighters began to pour out of the warships.

"Yes sir." Dan said, as they all watched spellbound as seventy-two shiny silver Teardrop fighters materialized on the flats screens.

"Are you getting anything Addi?" Brent asked, keeping his eyes glued to the screens in front of him.

"I just found their frequency. They see your Teardrops and they think the space people are attacking them, and those missiles of yours knocked out a bunch of their fighters so they're in a panic."

"Good, that's what we were praying for." Brent said, as the enemy fighter's began to form up into a V shaped battle array.

"They are being ordered to pursue your fighters or face severe consequences." Addi said, standing up so he could see better, or because he was too excited to stay seated.

"There they go." Brent said, stating the obvious as everyone watched the small black Snark fighters take off after Dan and his strike team.

"Is Peter ready?" Toby asked, thinking that the second strike force should be in position by now.

"Pete, are you ready?" Brent asked, speaking into his microphone.

"Yes, we're all in position. Whatever comes out of those warships next is dead meat!" Peter said, just as a black five-man scout ship emerged.

"Let them both come out before you fire!" Brent yelled into his microphone just as the second ship emerged and one of Peter's teams fired. Again, they couldn't see the explosions taking place inside the enemy scout ships, but it quickly became obvious that they were out of action as one of them careened into it's mother ship and bounced off.

"Their Commanders are screaming at them, but they aren't making sense. Some of them are yelling at them to make a run for it and some of them are telling them to get back inside." Addi said, as scout ship after scout ship emerged and was hit by missiles that seemed to come at them from empty space.

"That's it, we got them all." Peter said, his voice coming from the big speakers.

"Then get out of there, make sure none of the fighters get back to their mother ships." Brent said, ordering Peter and his fighters away from the Snark warships.

"There go the black widows." Toby said, as

three small black dots materialized out of thin air and headed for each of the monstrous warships. "God have mercy on their souls, because those things won't." He added, shaking his head as they hit and immediately extended their eight legs and began to scramble across the hull; hunting for warm bodies.

"Look!" Neli said, standing up next to his uncle. "One of them just went into that ship. I thought they were supposed to stay on the hull?"

"Where?" Brent asked, not seeing the one Neli was talking about.

"There." Neli said, leaning over Brent's shoulder and pointing at one of the warships on the screen. "One of your black widows just went in through that force field and into the hanger area."

"Is that going to be a problem?" Addi asked, looking at the warship Neli had pointed out.

"If we can't see that it's shut down, we may have to destroy it." Brent said, thinking that's probably what they would have to do. "Addi, see if you can pick up the fighters. They are our biggest threat now."

"We just cloaked and fired all of our rear missiles, we'll circle around and join Hacker in laying down his broadside." Dan's voice said, coming out of the big speakers as clear as if he was standing there and talking to them in person.

"Look!" Toby yelled, as all of the operators started putting up camera shots from Hacker's group. "You can see the missiles they fired, they're smashing them to bits." He added, as everyone watched as missile after missile hit one of the Snark fighters.

"Addi?" Brent said, calling out Addi's name as

a question.

"They're mostly screaming. They can't figure out who's firing at them or what happened to the ships they were chasing."

"Hacker's given the order to fire." Brent said, as thousands of missiles suddenly appeared from the Snark fighter's left and began the grisly business of slaughtering the black fighters by the hundreds."

"Tell Hacker to watch out, someone just ordered all of them to fire everything they have at the area of space where the missiles are coming from and then to make a run for it!" Addi yelled out, standing to his feet as all of the remaining Snark fighters turned and fired everything they had at Hacker's fighters.

"Hacker, get out of there!" Brent screamed into his microphone, knowing it was too late as two orange blossoms of fire erupted in the area where Hacker had been firing from and then winked out.

"They're making a run for it." Toby said, watching as the remaining Snark fighters turned and made a beeline back to their mother ships. Only to start exploding as they ran into Peter's strike force.

"Did any of them make it back?"

"No." K'Narf said, speaking for the first time. "It's over, you and your team did a fantastic job Mister Conrad, now you need to spread the word, a lot of people have been praying for you."

"Thank you." Brent said, shaking K'Narf's offered hand.

"Who did we lose?" K'Narf asked, letting go.

"David Ben Ryan and his bond mate Jennifer and Thomas Michael Neely and his bond mate Janet." Brent said, having just gotten the information from Hacker.

"We need to have a council meeting and I want it held in Redwing's commissary. Notify Premo and Sylvia and Minh Le and tell Minh Le to bring all of his recording equipment. If they start now they will be able to get here by Sabbath. We can have the meeting Sunday morning after everyone's had a chance to rest up." K'Narf said, standing up.

"In the commissary?" Karl asked, thinking of how many people that would want to attend.

"Yes, haul out all of those round tables and move back the serving line and stuff the place full of those folding chairs of yours. You might want to hang flat screens outside for those that can't get inside." K'Narf said, beginning to see in his mind's eye what he wanted done.

"Yes sir." Karl said, already planning how he was going do it.

"Can you jury-rig up a podium that will hold about two dozen chairs so they can see us?" K'Narf asked, knowing that Karl knew he meant the whole council. "And a microphone system, we'll need one of those too."

"Yes, that should not be a problem." Karl said, thinking that he would have to have the podium pre-built so that all he had to do was have his guys haul it in at the same time they brought in the chairs.

"Oh, and I don't know where Debbie is, but make sure she gets here." K'Narf said, turning to look at Brent. "I believe you have a couple of pictures you will need to give her." K'Narf said, thinking of the two lost fighter teams.

"Yes sir." Brent said, quietly thanking the Lord for the precautions he had taken; knowing that

Debbie would have a studio quality picture to paint from.

"Then, unless you have questions, I want to go and meet my new son." K'Narf said, smiling at Brent and Toby.

"Son?" Toby said, his word a question.

"You two sure don't keep up with what's going on around here do you?"

"A son?" Toby asked again.

Yes, and I very much want to go home and see my wife and meet my new son . "Do you have someone that can take me?" K'Narf asked, looking at Brent.

"I'll take you myself." Brent said, getting up and pointing at the exit.

"I want to go too." Toby said, taking off after them, leaving Addi, Neli and Karl to fend for themselves.

Chapter Twenty-Six

"Thank you mother." K'Narf said, meaning it as he looked fondly at his mother.

"That's what grandmothers are for." Robbie said, rocking little Leecal in her arms. "Now you two run along to your big meeting or whatever it is. We'll be just fine, won't we?" She cooed, rocking back and forth in the new rocking chair she had Karl get for her.

"We shouldn't be gone too long." T.J. said, coming out of the bedroom looking like a million dollars in a white pants suit with dark green embroidery work on the sleeves and collar, her tummy so flat and smooth that it was impossible to tell that she had just had a baby. "We can't thank you enough." She said, adding her thanks to K'Narf's.

"Posh, go on with you now, he won't go to sleep with the two of you hanging around." Robbie said, trying to shoo them out with her eyes.

"You ready?"

"I guess so, you know that there are going to be a lot of people there tonight." K'Narf said, opening the door.

"Yes, I expect that there will be, so we might as well get it over with. Have you thought of what you want to say?"

"Kind of, it's not very clear in my mind yet."

"Well, let's hope your mind clears up before we get there." T.J. said, taking K'Narf's arm and heading toward the walkway and Redwing's commissary.

"Ah, there you are." Karl said, having waited outside for them. "We're going to have to go around to the back and enter through the compactor door. There's no way you're going to get through that crowd in there." He added, as he led them around the side of the building to a door that opened out next to a large trash compactor.

"Seems appropriate." K'Narf said, as Karl got out his keys and opened the door.

"Speak for yourself." T.J. said, not caring for K'Narf's inference.

"Celebrities are always smuggled in and out of buildings." K'Narf said, trying to undo what he had said as they wormed their way through the kitchen.

"That's not what you meant and you know it." T.J. said, refusing to let K'Narf off the hook as Karl ushered them up a makeshift stairway and through some jury-rigged curtains. An audience of thousands stared at them as they took their seats.

"Looks like everyone is here." K'Narf flashed, looking at all of the various members of what he loosely called his council, including Brent's lead pilots Dan, Peter, Hacker and their wives.

"Friends and neighbors of Etaguf City, I give you Mister K'Narf Etaguf!" Karl said, opening the meeting as he stepped up to a microphone and made his initial introduction. Turning he motioned for K'Narf to come up to the microphone; adjusting it up a few inches to accommodate K'Narf's height.

"Good morning. By now you've all heard the

news. We won!" K'Narf said, pausing as everyone both on and off the stage broke into applause.

"It's difficult to single out those most responsible for our victory, recognizing the long hours some of you spent assembling fighters or sweating over hot stoves so others could eat and get back to work, but there are some exceptions.

There are some that gave their all and I'd like to honor them by mentioning their names. The first name I'd like to mention very few of you probably know, but he lost his life helping put the dome over us. His name is William 'Little Willie' Getz and he has the dubious honor of being the first person buried in the Etaguf City Cemetery."

"Thank You." T.J. flashed, sending K'Narf a little red heart with "I Love You" written on it.

"Next are four names that you should all know, but just in case you haven't heard who they were, I would like to call out their names. These are the fighter teams that lost their lives defending this city and everyone in it and everyone living back on Earth. First there's David Ben Ryan, a young man I met when working down in the Sutton Wells factory. A young man I will miss very much. I only met his bond mate Jennifer once just prior to the battle, but she must have been an exceptional woman for David to have bonded with her. Next are a couple that I again had the pleasure of meeting only once just prior to the battle, Thomas Michael Neely and his bond mate and copilot Janet.

Debbie." K'Narf said, turning and nodding at Debbie. "Has been given, or will be given, two studio quality pictures of both of these lost teams so she can paint their portraits which will be hung alongside Wonders in the Crystal Chapel down in the Whale. In the coming years, whenever you have a chance to stop by and visit the Chapel, it will give you a chance to remember them and thank our Lord for their ultimate sacrifice.

Now there are other things we need to talk about." K'Narf said, the tone of his voice telling everyone that the time for victory and mourning had been set aside.

"Here we go again." Brent said, just loud enough for those on the stage to hear.

"T.J., Etaguf City's financial officer and her committee have, as you know, established a banking system. From what I have been made to understand it is working quite well, and I would like to thank her and her committee for their planning and foresight in setting it up." K'Narf said, turning and pointing at T.J.

"I'll say it is, I have people volunteering to dig foundations just so they can run up their credits." Karl said, shouting it out so everyone could hear him, getting cheers and applause as he said it.

"Thank you Karl, I'm glad you brought up the subject of earning credits, because there is an awful lot of work that still needs to be done and many positions that will need to be permanently filled."

"Permanently filled?" Karl asked, K'Narf's words catching him by surprise.

"Yes, starting with a full time standing military. Admiral Conrad." K'Narf said, turning and nodding at Brent so everyone would know exactly who he was referring to with the new title of Admiral. "Will be recruiting a force capable of manning his new fleet around the clock. I suggest something like the Blue and Gold teams the U.S. uses for their atomic submarine service, though you might consider a smaller rotation time cycle, say one month on duty and three months off which would require four teams instead of two though."

"Yes sir." Brent said, not knowing what else to say, surprised by his new title.

"And there will be millions of credits to be earned salvaging that Snark fleet and turning it into state-of-the-art fighting machines." K'Narf said, turning his attention back to his audience. "Not to mention thousands of permanent assignments within the military itself, all of which Admiral Conrad will be overseeing starting right after we adjourn this meeting." He added, knowing he had just set Brent up big time.

"There will also be considerable credits to be earned manufacturing the thousands of new Teardrop fighters that will be needed, and manufacturing these fighters, as with everything built within the walls of Etaguf City will be overseen by our ever popular new Etaguf City Manager, Karl Grauer." K'Narf said, as applause broke out at the announcement of Karl's new position and title.

"You did that on purpose so he couldn't run against you, didn't you?" T.J. flashed, knowing that K'Narf had done it on purpose.

"I don't know what you're talking about." K'Narf flashed back, as he continued.

"Now we need to talk about visiting family members that are not glow heads, like my mother, who is home babysitting Leecal, the first, but definitely not the last, baby to be born in Etaguf City." K'Narf said, as applause again broke out forcing him to pause for a moment before continuing.

"And for those of you who don't know them,

Addi and Zithri are going to be starting a business training mentors." K'Narf said, turning around and pointing at Addi and Zithri. "They both have the equivalent of a PhD. in child rearing, specifically glow head children. Zithri specializes in male pre-birth up through five years along with his brother who also specializes in male mentoring up through five years. It is my understanding, correct me if I'm wrong, but you will be training teams of mentors for mentoring females from pre-birth up through five years as well as teams to take the child from five years up through the age of twenty, and then other teams to take the child up to the age of one hundred where he or she will be considered an adult. Is that correct?" He asked, directing his question at Addi and Zithri.

"Yes, that is correct." Addi said, nodding at the audience, obviously nervous about being in front of such a large crowd.

"And if you're wondering who these men are and where they came from." K'Narf said, turning and pointing at Zithri, Addi and Neli. "These are the men the Holy Spirit sent us out looking for when we realized that we were going to be fighting an enemy we knew nothing about. Without their help it's possible that we wouldn't be here celebrating tonight.

They're the three men mentioned in the Bible that brought gifts to our Lord just after he was born, also known as the three wise men." K'Narf said, as everyone in the place stood and gave a resounding round of applause.

"And as for Neli, don't be fooled by his youthful looks or by the fact that his uncles refer to him as their *young* nephew." K'Narf said, stressing the word *young* as he turned and pointed at Neli. "When our Lord lived and walked the dusty roads of Israel, He called him Yochanan; however the Bible we know refers to him as John the apostle that our Lord loved." He added, as the whole room again erupted in a standing ovation.

"So remember, even though he acts like he's no older than his new best friend Brent Conrad, or should I say Admiral Conrad, he is actually your senior by several thousand years." K'Narf said, unable to stop the standing ovation as the whole room erupted again.

"If a child is not considered an adult until he is a hundred years old, I wonder what that makes us?" K'Narf asked, after everyone finally stopped applauding and sat back down.

"I don't know what it makes you, but I qualify." Karl said, talking low enough so only K'Narf and those on the stage could hear him.

"To bad age isn't synonymous with wisdom." K'Narf said, turning away from the microphone so the audience could not hear his response.

"What do you mean?" Karl asked, as K'Narf turned back to the microphone.

"Karl, did you give my mother an apartment that she plans to use when she and my father return to spend the New Year's Eve with T.J. and I?"

"Yes I did, why?" Karl asked, knowing instinctively that K'Narf was about to set him up.

"It was a nice gesture Karl and I appreciate it, but I don't think we're going to have enough apartments to go around if we give all our visitors an apartment of their own. I would like to suggest that you select a group of apartments and turn them into what they call time-shares, starting with the one you promised my mother, and I'll let you tell her of your change of heart." K'Narf said, seeing nods of approval in the audience as he turned to look at Karl to see if he agreed.

"Yes sir." Karl said, knowing K'Narf was right as K'Narf looked at him, but not wanting to admit it in front of everyone, and definitely not wanting to tell Robbie that she was going to have to share her new apartment.

"Understand!" K'Narf said, raising his voice as he turned his attention back to his audience. "This means that you will have to schedule any visit by your parents or other blood relatives through Karl. I say blood relatives because anyone other than a blood relative will have to be cleared specifically through Admiral Conrad's security service before visiting Etaguf City.

Something else you need to understand, there will be no pictures taken that will give proof that this city exists. Your relatives can take all the pictures they want of you inside your apartments, but any picture taken outside must be pre-approved by your new City Manager, Karl Grauer. You also need to impress on your visiting relatives that should they try to tell the world that we exist, we will work behind the scenes to get them committed, and that they will find themselves spending time in a padded cell, is that clear?" K'Narf asked, seeing heads all over the room nodding yes.

"DeeDee, we're going to need a hospital and a pharmacy up here to accommodate the needs of our normal visitors. I suggest at least twenty beds all set up like an Intensive Care Unit. Could you round up people with the necessary skills to man such a facility on an emergency only basis?"

"Yes sir, I should be able to do that." DeeDee said, immediately thinking of friends she knew who had medical backgrounds.

"Karl, will there be any problems fitting a hospital into your city planning layout?"

"No sir, it won't be a problem at all." Karl said, thinking it a good idea and wondering why none of his planners had thought of it.

"Brent, what's the matter?" K'Narf asked, seeing a look of concern on Brent's face.

Chapter Twenty-Seven

"His plane just went down." Brent said, listening intently to the flash he was receiving.

"His Plane?"

"Our new outspoken President-elect's plane!"

"How, I mean where?" K'Narf asked, a tingling sensation enveloping his mind as he asked the question.

"Hang on." Brent said, listening intently before answering K'Narf's question. "It seems he was headed to D.C. to be inaugurated and they think his plane was at full throttle when something went wrong and it exploded."

"How? I mean how do they know it exploded?"

"They don't, just guessing I think." Brent said, listening intently.

"But why do they think it exploded?"

"Because of the way it disappeared off radar and now everyone's in a panic." Brent said, looking at K'Narf as his flash connection was broken.

"Panic?" T.J. asked, confused.

"Yes, if it blew up in mid-air at those speeds they won't find a piece big enough to identify." K'Narf said, his whole body tingling as the words *stumble and fall* echoed through his mind.

"Every ship and search plane the Navy can muster is heading for the area where it disappeared to look for him." Brent said, looking around the room as other people started to get the same message he had just received.

"You need to get every search team you can put together into D.C. and the surrounding areas and scour them for our people and get them out. You have less than forty-eight hours." K'Narf said, seeing the meaning of his words dawn on Brent.

"You really think it's started?" Brent asked, disbelief shading his question as he stood up.

"Yes, the first domino has definitely been knocked over," K'Narf said, looking Brent directly in the eye as he added, " The first of the three sixes *...he shall stumble and fall and not be found*. Every nerve in my body is telling me its started." He added, as he turned back to the microphone, noticing that almost everyone was getting flashed by someone they knew and spreading the news about what had just happened.

"Don't take off yet, I may need you here." K'Narf said, speaking over his shoulder at Brent as he surveyed the room.

"Yes sir." Brent said, setting back down in his seat.

"May I have your attention please?" K'Narf asked, his plea having very little affect on the excitement rippling across the room full of people. "Please! May I have your attention!" K'Narf said again, raising his voice till it boomed out of the speakers planted around the room and outside the building, this time bringing instant silence, even to the murmur coming from outside.

"Thank you, I understand that most of you are getting the same news we just received up here, that the new President-elect of the United States has just disappeared at sea in a downed plane. I'm sure you understand that this means that one of the final pages of history has just turned and that we are now living through Daniel eleven and verses nineteen and twenty. It can mean only one thing and that is that the much talked about end times are finally upon us, and we have some decisions to make.

DeeDee." K'Narf said, surprising everyone as he turned to face DeeDee. "I would like you to give Brent, I should say Admiral Conrad, the names and last known addresses of the eight couples that were in your Bible study group. The ones you told us were doing exactly what the Bible told them to do; stocking up food and medicines so they could ride out the final three and a half years of the antichrist's reign in hiding."

"Why?"

"If we can find them we can keep an eye on them and if something should befall them and they get caught, we might be able to help them."

"What do you mean befall them?" DeeDee asked, thinking of her friends.

"A hurricane, tornado, earthquake or a flood for example, might destroy their ability to hide. If they hold to their belief in the Lord and continue to refuse to give obeisance to whatever it is the antichrist is going to place in that new Jewish temple, they will be declared fundamental terrorists and arrested."

"But how can we help them if they've already been picked up?" DeeDee asked, speaking her question loud enough for the microphone to pick it up and letting everyone hear it.

"Daniel says that '*Now when they shall fall, they shall be holpen with a little help.*' Now in my book the *'little help'* is us, and if we can act quick enough we might be able to get to them before they get *processed out* as you put it." K'Narf said, emphasizing the words *"processed out"*. "Understand, the Bible specifically says '*when they fall'*, not before they fall."

"Then you really think it's possible?"

"Possible?"

"That they can successfully hide out for that long?" DeeDee asked, knowing that she and her friends had talked about that a lot.

"You mean for three and a half years?" K'Narf asked, looking at the worried expression on DeeDee's face.

"Yes."

"Let me put it this way, if Bin Laden, the worlds most wanted man, a man with a twenty-five-million dollar reward on his head can successfully hide out for six years within spitting distance of the authorities hunting him, your friends shouldn't have any trouble staying out of sight with no one actually looking for them, at least by name. Does that help?" K'Narf asked, seeing the relief in DeeDee's face.

"Yes, but how can you get them out If they've already been picked up, I..."

"Brent." K'Narf said, cutting DeeDee off as he turned his attention to Brent. "Can you train a team that would be capable of going into a police station or a fortified prison and rescuing someone that they were holding?"

"Yes, you might say we already have."

"What do you mean?"

"One of my guys was hit from behind and knocked out. By the time he came to he was locked up in a holding cell in a heavily fortified police station in Iraq." Brent said, noticing how quiet it had gotten. "We not only went in and got him out, but we erased all records of his being arrested."

"All records?"

"Yes, one of my guys hacked into their computer system and deleted the record of his arrest." Brent said, smiling at the look on K'Narf's face.

"Does that answer your question?" K'Narf asked, looking back at DeeDee.

"Yes sir." DeeDee said, looking with a new level of respect at Brent.

"Does everyone understand what was just said?" K'Narf asked, looking around the room. "We have the ability to rescue people that have been picked up for refusing to deny the Lord and give obeisance to the antichrist, but God has specifically said we are not to rescue anyone unless they have fallen. Do you understand what that means?"

"You need to explain it a little better." T.J. said, sitting directly behind K'Narf so the microphone did not pick up her voice.

"Let me put it in another light. Karl, how many people will this city hold once all the apartment buildings are completed?"

"Comfortably, just shy of a million, crammed in maybe another hundred thousand."

"Did you hear that, Etaguf city can hold about a million people comfortably and another hundred thousand crammed in once all the buildings are finished. Now subtract our people and you have about three quarters of a million people that we can rescue. That's not a lot of people when you think of the millions that the Bible says will deny the antichrist.

I want to make sure you understand this, God says we are not to rescue anyone, family included, *unless they have fallen*." K'Narf said, emphasizing the words '*unless they have fallen*'. "Which means that they have been picked up and are in danger of being '*processed out*' as DeeDee puts it." He added, looking at a room full of somber faces as he emphasized "*processed out*", thinking what a dramatic change from just a few minutes earlier when they were cheering the successful end of their battle with the Snark fleet.

"All of our families need to understand what is about to happen. They may have less than four years before the antichrist starts persecuting anyone who does not verbally worship him and his blasphemous image. That's not a long time to figure out where and how they can do what the Bible says they have to do to survive, and that is to hide. By '*hide*' I mean do what DeeDee and her friends did, stock up and prepare to shut their gates and stay inside where they won't run afoul of the antichrist's scanning systems."

"I think they understand now." T.J. said, flashing her words to K'Narf rather than speaking them.

"Good!" K'Narf said, answering T.J. verbally so everyone could hear, even though they hadn't heard what she had said. "Karl, how long will it take you to finish this city and get it ready for guests?"

"Polished up and shining like a new penny, about three years." Karl said, thinking of the massive amount of work that would need to be done.

"Does that include building Teardrop fighters for Admiral Conrad's new fleet?"

"How many are you going to need?" Karl asked, looking over at Brent.

"Six hundred and sixty-two to be exact." Brent said, remembering how many enemy fighters the Snark warships could hold and figuring that would be how many fighter cradles they would have in them.

"Yeah, if Toby can get us the parts we can do that too." Karl said, thinking of all the newbies that were pouring in on a daily basis, and how many of them wanted jobs that would earn them credits.

"Brent, can you get your new fleet fully operational in the same three year period?"

"Yes sir."

"Then I would like to see a detailed timeline of how your going to do it, with monthly milestones laid out sometime next week, is that possible?"

"Yes sir." Brent and Karl both said in unison.

"DeeDee, food supplies for a million people for at least ten years, organically grown where possible and if you can't get it organic at least make sure it doesn't come from genetically modified sources, can you do it?"

"That's what we've been doing, so yes we can do it." DeeDee said, looking at Karl and Brent, thinking they'd got the worst end of the deal this time.

"Good, you do understand that the black horse of the apocalypse is still riding hard, and shortages and greed will drive food prices through the roof so you need to get futures written on as much as you can as quickly as you can." K'Narf said, looking at DeeDee.

"Yes sir." DeeDee said, looking over at Karl as she agreed, knowing that he would help her with the necessary contracts.

"Does anyone have any questions about what has happened, and what it is that we and the world our families and loved ones call home are getting ready to live through?" K'Narf asked, looking at every member of his council and then at the thousands of glow heads crowded into Redwing's commissary.

"Good, Minh Le, can you get copies of this over to Premo and Sylvia as soon as possible."

"Yes sir." Minh Le said, looking down the line of chairs and nodding at Premo and Sylvia.

"Then if there are no more questions, I'd like to have a few moments for silent prayer, we have a lot to do and a lot to pray about, but right now let's just thank the Lord for blessing and loving us as a people that have dedicated themselves to serving Him." K'Narf said, as he bowed his head.

Moments later he raised his head and said. "This council is adjourned."

"Everyone pick-up a chair on your way out, I want to clean the floor before we bring the tables back in. Chop, chop I'm hungry." Karl yelled into the microphone.

"Give a man a new title and first thing he does is start yelling orders."

"Who said that?"

"You were wonderful sweetheart, you're going to make an excellent mayor and your son will grow up and be so proud of you." T.J. said, taking K'Narf's hand as he helped her up.

"And what about my daughter?"

"Don't even think it!" T.J. cried, hitting K'Narf on the arm as hard as she could, which wasn't very hard as the hubbub of life broke out in every direction around them.

# # #

Other works by
Steven A. Vaughn

<u>Non-Fiction</u>

Coming Prophecies!!!
My Mother's Prayers
Spiritual Pathways

<u>Short Fiction</u>

God's Gift Giving Day

<u>Novellas</u>

The Missing Link Church
We Cannot Afford

<u>Full Length Novels</u>

Series Reading Order

Book 1 - Temani: Odd Angel Out
Book 2 - Temani: Odd Angel Out II
Book 3 - K'Narf Etaguf
Book 4 - K'Narf Etaguf II
Book 5 - Holpen
Book 6 - Holpen II
Book 7 - Holpen III

Written for Christians by a Christian, please tell a friend!

Made in the USA
Monee, IL
04 November 2026

67386440R00193